The Dorling Kindersley
BIG BOOK OF
KNOWLEDGE

DK

A DORLING KINDERSLEY BOOK
Conceived, edited, and designed by DK Direct Limited

General Editor Sarah Phillips

General Art Editors Ruth Shane, Paul Wilkinson

Subject Editors Francesca Baines, Val Burton, Deborah Chancellor,
Roz Fishel, Hilary Hockman, Rosemary McCormick,
Sarah Miller, Seán O'Connell, Jean Rustean, Anne de Verteuil

Subject Art Editors Flora Awolaja, Liz Black, Amanda Carroll,
Richard Clemson, Ross George, Sarah Goodwin, Sara Nunan,
Ron Stobbart, Andrew Walker, Sonia Whillock

Designers Wayne Blades, Alison Greenhalgh, James Hunter,
Marianne Markham, Tuong Nguyen, Alison Verity, Samantha Webb

U.S. Editor B. Alison Weir

Picture Researchers Lorna Ainger, Caroline Brook, Veneta Bullen,
Anna Lord, Miriam Sharland, Paul Snelgrove

Subject Consultants Dr. Michael Benton, Nicholas Booth,
Martyn Bramwell, Roger Bridgman, Eryl Davies, Liza Dibble,
John Farndon, Adrian Gilbert, Alan Harris, Paul Hillyard,
Dr. Malcolm MacGarvin, Diana Maine, Margaret Mulvihill,
Dr. Fiona Payne, Chris Pellant, Dr. Richard Walker, Shane Winser

Production Manager Ian Paton
Production Assistant Harriet Maxwell

Editorial Director Jonathan Reed
Design Director Ed Day

First American Edition, 1994
Published in the United States by
Dorling Kindersley Publishing, Inc., 95 Madison Avenue
New York, New York 10016

Copyright © 1994 Dorling Kindersley Limited, London

Library of Congress Cataloging-in-Publication Data
The Dorling Kindersley big book of knowledge. — 1st American ed.
p. cm. — (Picturepedia)
Includes index.
ISBN 1-56458-518-2
1. Science—Miscellanea—Juvenile literature. 2. Technology—
Miscellanea—Juvenile literature. 3. Natural history—Miscellanea
—Juvenile literature. 4. Social sciences—Miscellanea—Juvenile
literature. [1. Science—Miscellanea. 2. Technology—Miscellanea.
3. Natural history—Miscellanea.] I. Title: Big book of knowledge.
Q163.D67 1994
500—dc20 92-54484
CIP
AC

This edition published by
Reader's Digest Young Families, Inc., with the permission
of Dorling Kindersley Publishing, Inc.

The Dorling Kindersley
BIG BOOK OF
KNOWLEDGE

DK

DORLING KINDERSLEY

LONDON • NEW YORK
STUTTGART

CONTENTS

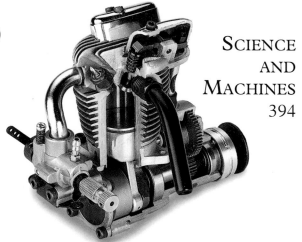

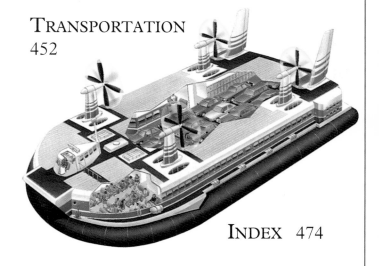

How to Use This Book

It's easy to use *The Big Book of Knowledge*. Start by looking at the contents page. There you will see that the book has four chapters, which are divided into different subject areas. Find a subject or a picture that interests you and turn to that page. If you can't find what you want on the contents page, look it up in the index at the back. This is an alphabetical list of everything covered in the book. After each entry, there is a list of all the pages on which that topic is mentioned.

See how the index works by looking up "dinosaurs." Turn to all the pages listed after the entry until you find the section "Enormous Meat-Eaters," shown here.

This page uses a picture of that section to tell you about some of the regular features you will find in *The Big Book of Knowledge*.

Enormous Meat-Eaters

Like tigers today, Tyrannosaurus rex may have hunted alone, terrorizing its prey with surprise attacks. This fierce dinosaur was a carnivore, which means it lived on a diet of meat. The biggest meat-eater ever to walk on this planet, Tyrannosaurus rex was heavier than an elephant and as tall as a two-story building. Its name means "king of the tyrant reptiles."

Hunters need good eyesight and a strong sense of smell. Large parts of Tyrannosaurus brain controlled its sense of sight and smell.

Tyrannosaurus had a massive skull. It took the shock of crashing into prey at speeds of up to 20 miles (32 kilometers) per hour.

Tyrannosaurus may have charged with open jaws, ready to sink its deadly teeth into its prey. Big chunks of flesh were swallowed whole.

Its short "arms" weren't long enough to reach its mouth, but they may have been used to grip and kill prey.

Duck-Billed Dinner
Tyrannosaurus probably hunted duck-billed dinosaurs, called hadrosaurs. It may have hidden among the trees, waiting for the right moment to charge at a peaceful herd of grazing duckbills.

Tyrannosaurus had a massive skull. It took the shock of crashing into prey at speeds of up to 20 miles (32 kilometers) per hour.

Captions
Most double pages feature one large, exciting picture. All around it, captions point out important details. They will help you look carefully at the picture and understand it.

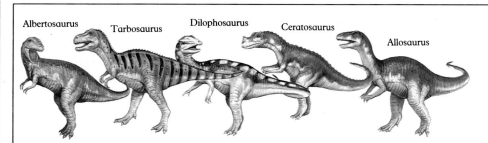

Albertosaurus Tarbosaurus Dilophosaurus Ceratosaurus Allosaurus

Picture Catalogs
On some pages, you will find a row of small pictures that are similar to the large one. Try to figure out the differences between them.

Enormous Meat-Eaters

Like tigers today, Tyrannosaurus rex may have hunted alone, terrorizing its prey with surprise attacks. This fierce dinosaur was a carnivore, which means it

*Hunters n
eyesight a.
sense of si.
parts of T
brain cont*

Title and Introduction
Each double page deals with a different subject. The title tells you exactly what the subject is, and the introduction gives you some basic information about it.

Step by Step
Occasionally, you will find a series of illustrations that shows you how something works. When you look at these pictures, for instance, you can see that Tyrannosaurus is using its small arms to help it stand up!

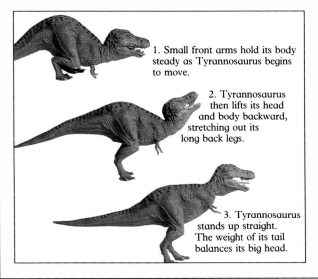

1. Small front arms hold its body steady as Tyrannosaurus begins to move.

2. Tyrannosaurus then lifts its head and body backward, stretching out its long back legs.

3. Tyrannosaurus stands up straight. The weight of its tail balances its big head.

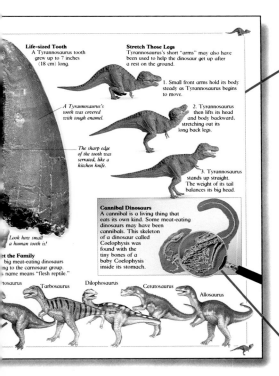

Life-sized Tooth
A Tyrannosaurus tooth grew up to 7 inches (18 cm) long.

A Tyrannosaurus's tooth was covered with tough enamel.

The sharp edge of the tooth was serrated, like a kitchen knife.

Look how small a human tooth is!

Stretch Those Legs
Tyrannosaurus's short "arms" may also have been used to help the dinosaur get up after a rest on the ground.

1. Small front arms hold its body steady as Tyrannosaurus begins to move.

2. Tyrannosaurus then lifts its head and body backward, stretching out its long back legs.

3. Tyrannosaurus stands up straight. The weight of its tail balances its big head.

Cannibal Dinosaurs
A cannibal is a living thing that eats its own kind. Some meat-eating dinosaurs may have been cannibals. This skeleton of a dinosaur called Coelophysis was found with the tiny bones of a baby Coelophysis inside its stomach.

et the Family
big meat-eating dinosaurs
ng to the carnosaur group.
name means "flesh reptile."

...osaurus   Tarbosaurus   Dilophosaurus   Ceratosaurus   Allosaurus

Measurements
In most cases, units of measurement are spelled out, but in some places you will come across the following abbreviations.

cm	=	centimeters
m	=	meters
km	=	kilometers
km/h	=	kilometers per hour
in.	=	inches
ft.	=	feet
m/h	=	miles per hour

Cannibal Dinosaurs
A cannibal is a living thing that eats its own kind. Some meat-eating dinosaurs may have been cannibals. This skeleton of a dinosaur called Coelophysis was found with the tiny bones of a baby Coelophysis inside its stomach.

Stories in a Box
Amazing facts or stories appear in a box. Sometimes boxes suggest experiments to carry out or things to do that will help you understand the subject better.

CHAPTER 1

EARTH AND SPACE

To us, our planet Earth seems enormous, but if we were able to gaze at it across the vastness of space, it would look like a tiny speck. It is one of the nine planets that are constantly hurtling around a star – our Sun – along individual oval paths, called orbits.

Together, the Sun and its planets are known as the Solar System. This, in turn, is part of a cluster of millions of stars and planets, called a galaxy. Our galaxy, which is shaped like a spiral, is called the Milky Way. It is so huge that a jet would take 100 trillion years to fly across it. So far, we have discovered about 6 billion different galaxies, and together they make up the universe.

Stars are made from layers of burning gas around a dense core. Some planets are mostly gas, too, while others are liquid. Our Earth, however, is like a gigantic ball of rock.

The Earth
Space

THE EARTH

Imagine you are an astronaut looking at the Earth from your spacecraft. What you see is a big blue ball covered with swirling clouds that hide features such as continents and mountains. The ball looks so blue because more than two-thirds of it is covered with water in the form of oceans, seas, lakes, and rivers.

The surface of the Earth, called the crust, moves all the time, but this movement is so slow that we are not aware of it. Eventually, however, pressure builds up and causes earthquakes and volcanoes. Changes also happen when the crust is worn away by water or huge blocks of moving ice, called glaciers.

Our planet can support life only because it gets light and heat from the Sun. Without it, the Earth would be a cold, dark, and dead place.

The Himalaya mountains in Nepal look very different from far out in space.

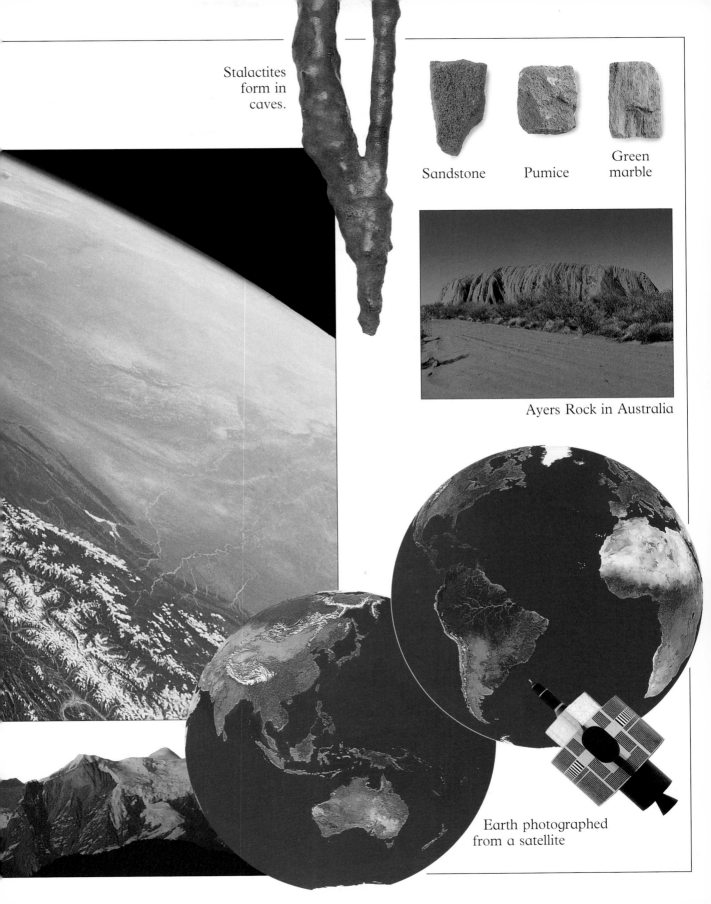

Stalactites
form in
caves.

Sandstone

Pumice

Green
marble

Ayers Rock in Australia

Earth photographed
from a satellite

THE EARTH'S CRUST

Just like you, the Earth has a very delicate skin. It is so thin that if you compare it to the whole Earth, it is thinner than the skin of an apple.

The Earth's skin, or crust, is made up of rock, built up in layers over millions of years. The layers look like blankets on a bed, with lots of lumps and bumps in them.

The crust is a thin layer of rock between 3½ and 42 miles (5.6 and 67.6 km) thick.

The mantle is the layer below the crust. In parts of it, the rock has melted like butter.

The outer core is made of iron and nickel that have melted to form a liquid.

The inner core is a ball of iron and nickel. It is hotter here than at the outer core, but the ball stays solid.

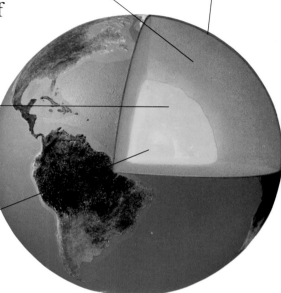

How Mountains Are Made

Mountains are made when the Earth's crust is pushed up in big folds or forced up or down in blocks. The different shapes made are given different names.

Overfold

Downfold

Upfold

The ocean is on top of the oceanic crust. The oceanic crust also runs underneath the continental crust.

Land is made out of the continental crust. It is thickest where mountains are found.

Under the oceans, the crust is as little as 3½ miles (5.6 km) thick, but under the continents, it is up to 42 miles (67.6 km) thick.

The mantle

Block mountain

Fault

Rift valley

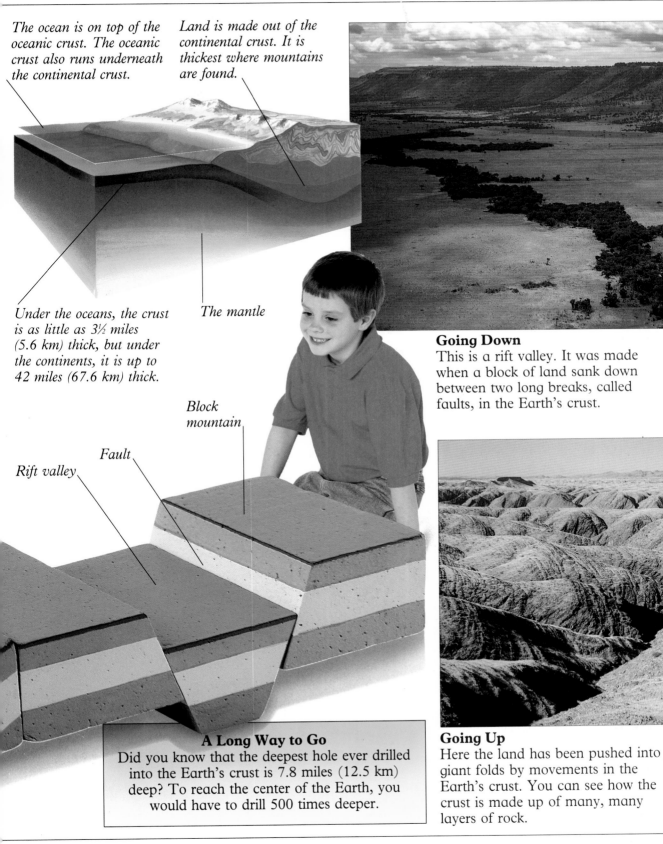

Going Down
This is a rift valley. It was made when a block of land sank down between two long breaks, called faults, in the Earth's crust.

A Long Way to Go
Did you know that the deepest hole ever drilled into the Earth's crust is 7.8 miles (12.5 km) deep? To reach the center of the Earth, you would have to drill 500 times deeper.

Going Up
Here the land has been pushed into giant folds by movements in the Earth's crust. You can see how the crust is made up of many, many layers of rock.

MOVING PLATES

The Earth's crust is not one unbroken piece. It is made up of many pieces that fit together like a giant jigsaw puzzle. These pieces, called plates, ride on soft, partly melted rock moving underneath them. The pieces push against each other, creating spectacular effects: earthquakes split the crust, volcanoes form, new land is made, and huge mountain ranges are pushed skyward.

All Scrunched Up
Sometimes, two plates push against each other and crumple the land, making huge mountain ranges.

Going Down
Sometimes, one plate slides under another. It is pushed down into the mantle and melts.

Doing a Split
Sometimes, two plates split apart, and lava bubbles up to fill the gap. It hardens and forms new land.

Slip-Sliding Away
Sometimes, two plates slip sideways past each other. This kind of movement causes earthquakes.

The red dots show you the places where volcanoes erupt.

Continent

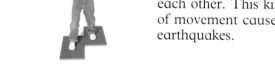

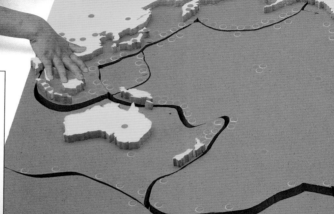

On the Move
The plates move all the time. In one year, they will usually move about one inch (2.5 cm). That's about as much as your fingernails grow in the same amount of time.

Past, Present, Future

Have you ever wondered what the Earth looked like in the past? These pictures show you how the continents have moved over the last 300 million years, and how the world may look 50 million years from now.

300 MILLION YEARS AGO

Changing Places

The land is coming together to make one gigantic continent.

200 MILLION YEARS AGO

PANGAEA

All Together

The super continent has come together. We call it Pangaea.

The Restless Earth

These houses and roads in Iceland are near a spot where two plates meet and have split the land.

150 MILLION YEARS AGO

LAURASIA

GONDWANALAND

Worlds Apart

The land is drifting apart again. Pangaea is splitting into two parts: Laurasia and Gondwanaland.

The green dots show you the places where earthquakes occur.

This is where two plates meet.

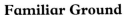

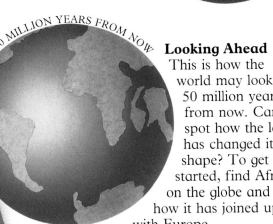

TODAY

Familiar Ground

Today, the world looks like this – but the continents are still moving.

50 MILLION YEARS FROM NOW

Looking Ahead

This is how the world may look 50 million years from now. Can you spot how the land has changed its shape? To get started, find Africa on the globe and see how it has joined up with Europe.

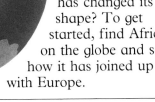

VOLCANOES

When you shake up a can of soda and then pull off the tab, the contents shoot out with a whoosh! A volcano acts a bit like this. With tremendous force, melted rock bursts through weak parts in the Earth's crust and is hurled high into the sky.

Volcanoes can be quiet and not erupt for a long time.

Hot springs are often found near volcanoes.

Nature's Fireworks
This volcano is putting on its own spectacular fireworks display. The explosions of red-hot lava and ash from the crater look like gigantic Roman candles.

The Spotter's Guide to Volcano Shapes

Spreading Out
Some volcanoes are flat. Their lava is very runny, so it spreads out in a thin sheet.

Short and Plump
Some volcanoes are squat. They are made of ash, which is lava that has turned to dust.

Going Up
Some volcanoes have pointed cones. Their lava is thick and sticky, so it does not run far.

River of Fire

The red-hot molten rock that is streaming down the sides of this volcano is beautiful but deadly. It is so hot that it can melt steel.

Clouds of ash and gas pour out from the crater.

Molten rock, called magma, rises up the main pipe and any branch pipes.

A volcano builds up from layers of ash and lava.

Branch pipe

Magma collects in a chamber found deep underground. It is forced up through cracks and holes in the ground.

Mixed Bunch

When lava cools and hardens, it can make rocks with different shapes. Here are three types:

Runny lava

Ropy lava

A volcanic "bomb"

EARTHQUAKES

Our planet is a restless place. Every 30 seconds, the ground suddenly rumbles and trembles. Most of the movements are so slight that they are not felt. Others bring complete disaster. Big cracks appear in the land, streets buckle, and buildings crumble. Whole towns and cities can be destroyed. Then everything settles down but is totally changed. The Earth has shaken. An earthquake has happened.

Fires are started by broken gas pipes and broken electrical cables.

Fallen telephone lines

Cars are smashed and they settle at crazy angles.

On this side of the fault the land has moved toward you.

Unsafe Ground
This is the San Andreas Fault in California. Earthquakes regularly happen here.

Terror from the Sea
Earthquakes under the ocean can cause giant, destructive waves called tsunamis, or tidal waves.

An earthquake occurs along a fault in the seabed.

Tsunamis can travel many miles across the ocean.

Why Earthquakes Happen

You may think that your feet are firmly on the ground, but the Earth's crust is moving all the time. It is made of moving parts called plates. When the plates slide past or into each other, the rocks jolt and send out shock waves.

Shaken Up

The Mercalli Scale measures how much the surface of the Earth shakes during an earthquake. There are 12 intensities, or grades. At intensity 1, the effects are not felt, but by intensity 12, the shock waves can be seen and there is total destruction.

Fault line

On this side of the fault the land has moved away from you.

What To Do in an Earthquake

Indoors, lie down under a bed or heavy table, or stand in a doorway or a corner of a room. After a minute, when the tremors will usually have finished, go outside, away from buildings, to a wide-open space.

Earthquake Words

The place within the Earth where an earthquake starts is called the focus. The earthquake is usually strongest at the epicenter. This is the point on the Earth's surface directly above the focus. The study of earthquakes and the shock waves they send out is called seismology.

Destructive Force

A tsunami piles up and gets very tall before it crashes onto the shore. It is so powerful that it can smash harbors and towns and sweep ships inland.

A tsunami can be almost 200 feet (61 m) high and can travel as fast as a jet.

ROCKS

Movements in the Earth's crust are slowly changing the rocks that make up the surface of our planet. Mountains are pushed up and weathered away. The fragments are moved and made into other rocks. These rocks may be dragged down into the mantle and melted by its fierce heat. When a volcano erupts, the melted rock is thrown to the surface as lava, which cools and hardens as rock. This is broken down by weathering, and so the cycle starts again.

Limestone

Conglomerate

Sedimentary Rocks

These are made from bits of rock and plant and animal remains. They are broken into fine pieces and carried by rivers into the sea. They pile up in layers and press together to make solid rock. The Painted Desert in Arizona is made of sedimentary rocks.

Red sandstone

In the Beginning

Rocks belong to three basic types. Igneous rocks are made from magma or lava and are also known as "fiery" rocks. Sedimentary rocks are made in layers from broken rocks. Metamorphic rocks can start off as any type. They are changed by heat and weight and are called "changed form" rocks.

In time, material moved by rivers and piled up in the ocean will become sedimentary rocks.

Rock fragments are carried from one place to another by rivers, glaciers, the wind, and the sea.

Recently formed sedimentary rocks

Igneous Rocks

These are made from magma or lava. It cools and hardens inside the Earth's crust or on the surface when it erupts from a volcano. Sugar Loaf Mountain in Brazil was once igneous rock under the crust. The rocks above and around it have been worn away.

Marble

Metamorphic Rocks

These are igneous or sedimentary rocks that are changed by underground heat, underground weight, or both. This marble was once limestone, a sedimentary rock. It was changed into marble by intense heat.

Slate

Obsidian   Granite

Surface rocks are broken down by the weather and by the scraping effect of tiny pieces of rock carried in the wind or in the ice of glaciers.

Glacier

Some rocks are thrust up as mountain ranges when the crust moves and makes giant folds.

Volcano

Lava that erupts from a volcano forms extrusive igneous rock.

Molten rock that cools and hardens inside the Earth is called intrusive igneous rock.

Metamorphic rocks, formed by heat and pressure

Folded rocks

Magma

CAVES

Caves are hollows beneath the surface of the Earth. The biggest ones are all found in rock called limestone, and some are huge. The world's biggest cave, in Sarawak in Borneo, is so large it could fit 800 tennis courts in it. Yet these caves began simply as cracks or holes in the rock that, over thousands of years, were made bigger by rainwater trickling into them and eating them away.

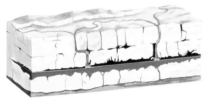

Drip . . .
The rainwater that seeps into the ground is very slightly acid and begins to eat away the limestone.

Drip . . .
The rainwater eats through the rock. It widens the cracks into pits, passages, and caves.

Drip
Over thousands of years, the passages and caves may join up to make a huge underground system.

Going Down
Water dripping from the ceiling of a cave leaves behind a mineral called calcite. Very slowly, this grows downward in an icicle shape that is called a stalactite.

The stream disappears underground into a pothole.

This pothole, or tunnel, leads straight down through the rock. It was made by a stream wearing away the rock.

Limestone is a very common rock. It is made from the skeletons and shells of tiny sea creatures that died millions of years ago.

Tunnel of Lava

Caves are found in rocks other than limestone. This one is made of lava and is inside a volcano in Hawaii.

Cracks in the rock are widened when rainwater seeps along them.

Limestone pavements are made when the rock is eaten away along joint lines.

Caving In

Sometimes a cave turns into a gorge. This happens when the roof falls in to reveal underground caverns and, far below, the river that carved them.

Cliff

Gallery

Cave mouth

Stream

Going Up

Where water drips onto the cave floor, columns of calcite, called stalagmites, grow upward.

OCEANS

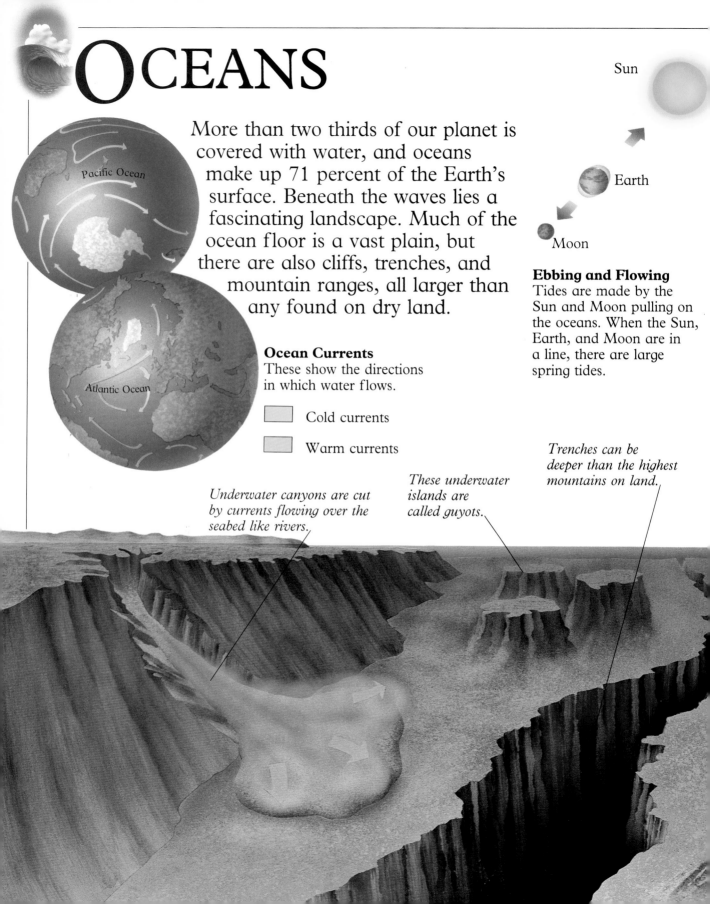

More than two thirds of our planet is covered with water, and oceans make up 71 percent of the Earth's surface. Beneath the waves lies a fascinating landscape. Much of the ocean floor is a vast plain, but there are also cliffs, trenches, and mountain ranges, all larger than any found on dry land.

Pacific Ocean

Atlantic Ocean

Ocean Currents
These show the directions in which water flows.

Cold currents

Warm currents

Sun

Earth

Moon

Ebbing and Flowing
Tides are made by the Sun and Moon pulling on the oceans. When the Sun, Earth, and Moon are in a line, there are large spring tides.

Trenches can be deeper than the highest mountains on land.

These underwater islands are called guyots.

Underwater canyons are cut by currents flowing over the seabed like rivers.

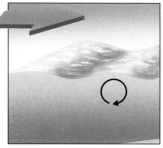

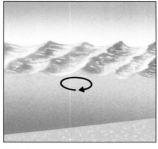

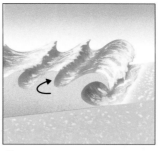

Surface

Lighted zone
660 feet
(200 m)

Dark zone
20,000 feet
(6,096 m)

Deepest zone,
a trench of
36,300 feet
(11,064 m)

Going . . .
The water inside a
wave moves around
and around in a circle.
It is the wind that drives
the wave forward.

Going . . .
Near the shore, the
circular shape of the
wave is changed, and it
becomes squashed.

Gone
The top of the wave
becomes unstable. When
it hits the beach, it
topples and spills over.

Ocean Currents
The direction in which currents
move depends on winds and the
Earth's spin. Winds blow the top
of the oceans forward, but the
Earth's spin makes the water
below go in a spiral.

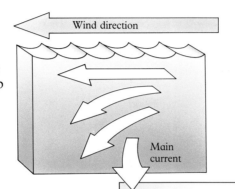

Wind direction

Main
current

The Dark Depths
Even in clear water,
sunlight cannot reach
very far. The oceans
become darker and
darker the farther down
you go – until everything
turns inky black.

*This island is a volcano
that has erupted from
the ocean floor.*

*A long, wide
ocean ridge*

*Water, heated by
the hot rocks, shoots
back into the sea.*

*Molten rock rises
up, cools, and
forms new
seabed.*

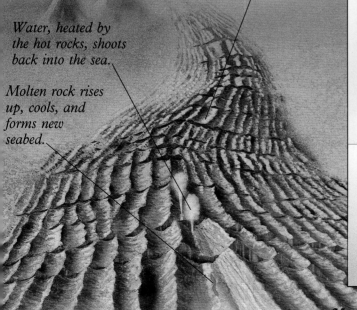

Frozen Worlds
In Antarctica and the
Arctic, the oceans freeze.
Icebergs break away from
glaciers flowing into the
water. Only a tiny part of
an iceberg is seen above the
surface of the ocean.

COASTLINES

Have you ever built a sand castle and then watched the sea come in, knock it down, and flatten it? This is what happens to the coastline, the place where the land and the sea meet. The coastline changes all the time because, every few seconds of every day, waves hit the land and either wear it away or build it up into different shapes.

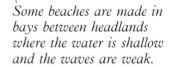

Some waves carry sand and pebbles from one area of the coast and leave them at another. This forms a new beach.

Headland

Going, Going, Gone
When caves made on both sides of a headland meet, an arch is formed. If the top of the arch falls down, a pillar of rock, called a stack, is left.

An arch

A stack

A cave forms over thousands of years as seawater creates cracks and holes in a cliff and makes them bigger.

Some beaches are made in bays between headlands where the water is shallow and the waves are weak.

Pounding Away
Waves pound the coastline like a giant hammer until huge chunks of rock are broken off. The chunks are then carried away by the sea and flung against the coastline somewhere else.

From Rocks to Sand
Waves roll rocks and boulders backward and forward on the shore. The boulders break into pebbles and then into tiny grains of sand. This change takes hundreds or thousands of years.

Shifting Sands
Dunes are made of sand blown into low hills by the wind.

An estuary is the place where a river flows into the sea.

Sea cliffs are one of the best places to see the different layers of rock.

Mud flats and marshes

Waves can build sand, mud, and pebbles into a long strip of new land, called a spit.

Living Rock
Coral is found in warm, sunny, shallow seas. It is made by tiny sea creatures that look like flowers. Over thousands of years, their skeletons build up into huge coral reefs and islands.

GLACIERS

A glacier is a huge river of ice that starts its life as a tiny snowflake. As more and more snow falls and builds up, in time it gets squashed under its own weight and turns to ice.

A glacier moves very slowly downhill. Because it is very heavy, it can push rock along like a bulldozer. It can wear away the sides of mountains, smooth off the jagged bits from rocks, and move giant boulders over tens of miles.

Mountains

Glaciers begin as huge snowfields.

The snow collects in hollows and turns to ice under its own weight.

Glaciers usually move downhill very slowly – no more than an inch or two (2.5 or 5 cm) a day.

The ice begins to move and rub away the sides and bottom of the hollow. Little by little, it changes the shape of the land and makes it into a U-shaped valley.

Close-Up View
The pilot in this plane is watching a wall of ice break away from a glacier and begin to crash into the water below.

Ice Power
When the water in this bottle freezes and turns to ice, it takes up more room and breaks the bottle. When the water that makes up the ice of a glacier freezes, it takes up more room and pushes away the rock.

Rubble is carried along by the glacier.

Shaping the Land
When you see a valley like this, you can tell from its U shape that it was once filled with the ice of a glacier.

Melted ice flows as streams and rivers inside most glaciers.

Bumps in the rock can be smoothed out by the ice moving downhill.

Out of Place
This giant boulder of hard rock was moved by a glacier and left on soft limestone. Then most of the limestone was weathered away, leaving a small block under the boulder.

Where a glacier flows into water, chunks of ice break off and float away.

Rocks carried along by the glacier pile up when the glacier starts to melt and stops pushing them.

The lower end of the glacier is called the "snout."

When the glacier melts, it makes new rivers.

RIVERS

Rivers are very powerful, so powerful that the force of the moving water is able to change the shape of the land. As they flow through mountains and over plains, rivers carry away huge amounts of rock, sand, and mud. They then dump it somewhere else, usually on riverbanks or in the ocean, to make new land.

A river usually begins in mountains or hills. Its water comes from rain or melted snow.

Where the rock is hard, the river makes rapids or waterfalls.

Glacier

Over the Top
When a river tumbles over the edge of a steep cliff or over a hard, rocky ledge, it is called a waterfall. This one is in Brazil.

As the river flows quickly down steep slopes, it wears away the rock to make a V-shaped valley.

Oxbow lake

Sand, mud, and gravel are left by the water as sediment.

Around the Bend
When a river reaches flat land, it slows down and begins to flow in large loops. It leaves behind sand, gravel, and mud, called deposits. This changes the river's shape and course.

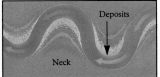

The river leaves deposits on the inside bend and eats away the outer bend.

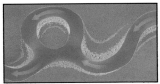

The deposits change the shape of the bend. In time, the neck of the bend narrows and the ends of the neck join up.

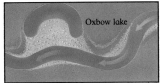

The river leaves behind a loop. It is called an oxbow lake because of its shape.

River Deep
A deep, narrow valley carved by a river is called a gorge. The Colorado River in the U.S.A. has made the world's largest gorge – the Grand Canyon. It is 1 mile (1.6 km) deep and 218 miles (350 km) long.

River's End
This swampy land is part of the Yukon Delta in Alaska.

A wide bend or meander goes across flat country.

As it reaches the sea, the river divides into small streams, leaving a mass of sand, mud, and rock fragments, called a delta.

Record Rivers
The longest river in the world is the Nile, in Africa. It is 4,160 miles (6,690 km) long. The largest delta covers 30,000 square miles (77,700 km²). It is made by the Ganges and Brahmaputra Rivers, in Bangladesh and India.

DESERTS

Did you know that deserts come in many different forms? They can be a sea of rolling sand, a huge area of flat and stony ground, or mountainous areas of shattered rock. There are hot deserts and cold deserts. So what do these very different areas have in common? The answer is that they are all very dry, and they all get less than 10 inches (25 cm) of rain each year. This rain may not fall regularly. Instead, it may all come in a single day and can cause a dramatic flash flood.

Sea of Sand
A desert may be hard to live in, but it can be stunning to look at. These dunes are in Saudi Arabia.

Tail dune

Wind

Barchan dunes

Seif dunes

Star dunes

Wind Power
Wind blows the sand into hills, which are called dunes. These have different shapes and names.

Cuesta

Mesa

Natural rock arch

Chimney or pipe rock

Butte

Dunes

On the Move

Imagine your hair dryer is the wind. It blows sand up the gentle slope of the dune. When the sand gets to the top, it tumbles down the steep slope. As more and more sand is moved from one slope to another, the whole dune moves forward.

Heavy rain causes flash floods. These rush over the land, loaded with sand and stones, and cut deep channels in the surface of the desert.

Broken rocks slide downhill and collect in gullies.

Water Power

The tremendous power of water has made this deep ravine near an oasis in Tunisia.

Where the rock is hard, ridges will stand out in the landscape.

Shaping the Land

Wind-borne sand blows against the rocks and wears them into beautiful and surprising shapes.

Steep slopes of broken rock

Outwash fan

Hot and Cold

This is one of the Devil's Marbles in the Northern Territory, Australia. The rock's outer layers have started to peel off because of the desert's very hot and very cold temperatures.

SPACE

The universe is made up of galaxies, stars, planets, moons, and other bodies scattered throughout space. A galaxy is a group of millions of stars. Our galaxy, which is shaped like a spiral, is called the Milky Way.

On a clear night, it is possible to see thousands of stars, which appear as twinkling points of light. Earth's moon is usually clear, and sometimes you can also see five of the planets: Mercury, Venus, Mars, Jupiter, and Saturn. These do not twinkle. They look like small, steady disks of light. Earth is the third planet from the Sun, which is about 93 million miles (149 million kilometers) away from us.

People have always been curious about the things they could see in the sky. It is only quite recently, though, that science has developed the advanced technology needed to send men and women into space.

Space shuttle

Milky Way

NEPTUNE

URANUS

SATURN

PLUTO

34

MERCURY

SUN

VENUS

EARTH

MARS

Asteroid belt

JUPITER

Rings

ROCKETS

Rockets were invented in China a long time ago. They looked a bit like arrows and worked by burning gunpowder that burned up very quickly, so the rockets did not travel very far. Since then, people have tried many ways of sending rockets up into space. In modern rockets, two liquid fuels are used. They mix together and burn. Then the hot gas shoots out of the tail, pushing the rocket up and away.

V-2 rocket 1945   Gemini Titan 1964

The Fly!
In 1931, a German named Johannes Winkler launched his HW-1 rocket. It went 7 feet (2 m) into the air, turned over, and fell back to the ground. A month later, he tried again. This time it climbed to 297 feet (90 m) and landed 660 feet (201 m) away.

3,2,1, Fire!
A hundred years ago, soldiers used rockets like this. They were called Congreve rockets.

Saturn Power
Saturn 5 is the biggest rocket ever built. It is as tall as a 30-story building! It was used in the Apollo program, which carried the first American astronauts to the Moon.

Fuel tank

See It Go!
If you blow up a balloon and let it go without tying a knot in the neck, the air will rush out very quickly. When the air goes out one way, it pushes the balloon the other way – just like a rocket!

The stabilizing fins keep the rocket on course.

Five rocket engines

Up, Up, . . .
How far can you throw a ball? About 50 or 60 feet (15 or 18 m)? It doesn't go on forever because the Earth's gravity pulls it back down again.

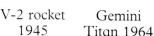

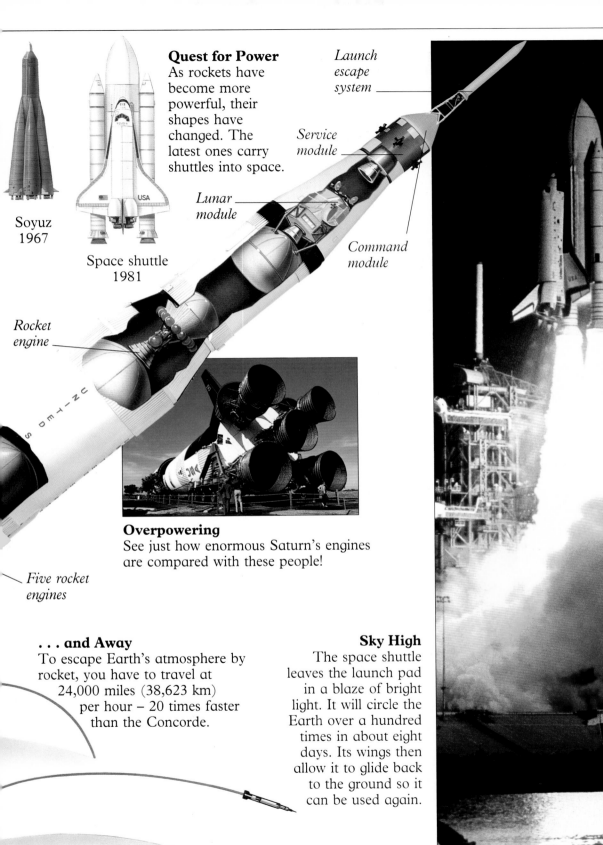

Quest for Power
As rockets have become more powerful, their shapes have changed. The latest ones carry shuttles into space.

Soyuz 1967

Space shuttle 1981

Launch escape system

Service module

Lunar module

Command module

Rocket engine

Five rocket engines

Overpowering
See just how enormous Saturn's engines are compared with these people!

. . . and Away
To escape Earth's atmosphere by rocket, you have to travel at 24,000 miles (38,623 km) per hour – 20 times faster than the Concorde.

Sky High
The space shuttle leaves the launch pad in a blaze of bright light. It will circle the Earth over a hundred times in about eight days. Its wings then allow it to glide back to the ground so it can be used again.

MOON MISSION

The second stage drops off when its five engines run out of fuel.

The command and service modules turn, join on to the lunar module, and pull it out of the third stage.

The Moon is the Earth's nearest neighbor in space, but it still takes three days to get there by rocket. It would take 200 days by car! When astronauts first went to the Moon, no one knew if it would be safe to land there. But American astronauts have been to the Moon on Apollo missions six times, and they all returned safely to Earth. The first Moon trip was in 1969 and the last in 1972.

Second stage

Astronauts traveling to the Moon crawl through a tunnel from the command module to the lunar module.

We Have Liftoff!
The first stage of the Saturn 5 rocket has five huge engines. When these run out of fuel, they fall back to Earth. Then the second stage takes over.

This air recycling unit keeps the air fresh in the cabin.

Lunar Module

CSM = Command and Service modules
LM = Lunar module
CM = Command module

CSM docks with LM

CSM in orbit

LM back to CSM

Moon

LM descends to Moon

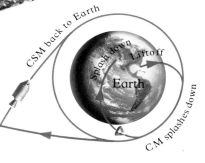

CSM back to Earth

Splash down

Liftoff

Earth

CM splashes down

Moon Trail
The Apollo mission to the Moon followed a path in the shape of the figure eight.

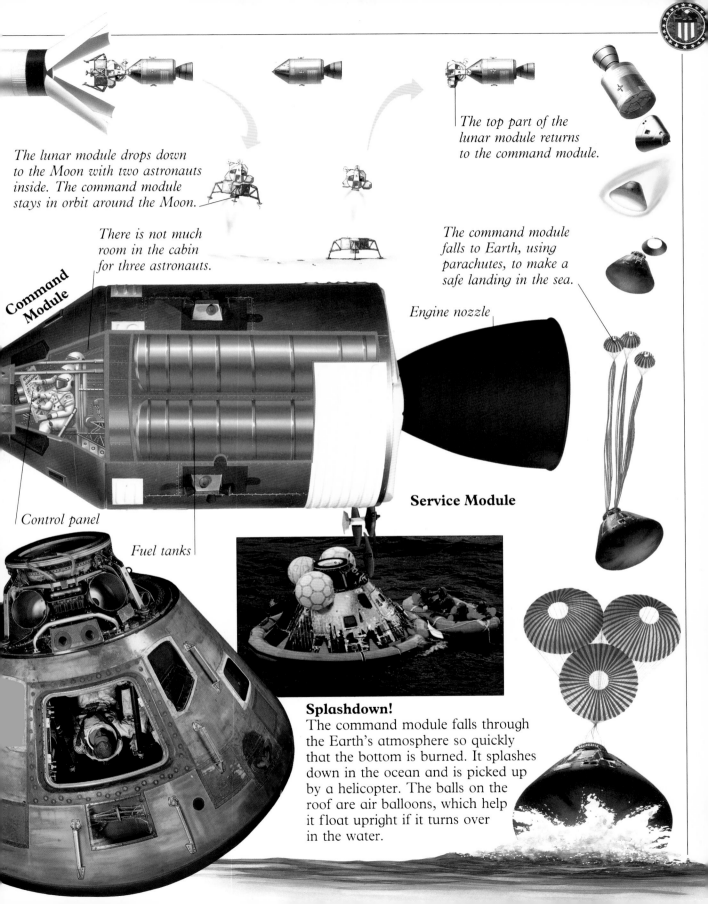

The lunar module drops down to the Moon with two astronauts inside. The command module stays in orbit around the Moon.

The top part of the lunar module returns to the command module.

There is not much room in the cabin for three astronauts.

The command module falls to Earth, using parachutes, to make a safe landing in the sea.

Command Module

Engine nozzle

Control panel

Fuel tanks

Service Module

Splashdown!
The command module falls through the Earth's atmosphere so quickly that the bottom is burned. It splashes down in the ocean and is picked up by a helicopter. The balls on the roof are air balloons, which help it float upright if it turns over in the water.

LUNAR LANDING

A lunar landing is a Moon landing. If you went to the Moon, you would find no living things at all, no air, and no water. If you stayed for a lunar "day"– about 28 Earth days – you would have two weeks of baking sun followed by two weeks of freezing nights. The first people on the Moon went down in the lunar module known as *Eagle*.

The *Apollo 11* Crew
Neil Armstrong and Edwin "Buzz" Aldrin were the first people to walk on the Moon. Michael Collins stayed in orbit in the command module.

Hanging Out the Wash?
No, just setting up a panel to collect dust! The Moon is covered in dusty soil and scattered rocks.

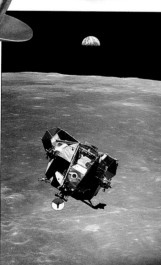

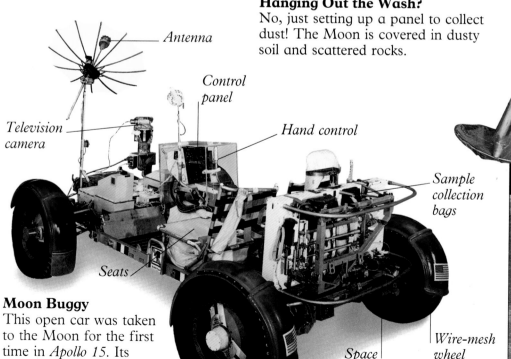

Antenna

Control panel

Television camera

Hand control

Sample collection bags

Seats

Space for storing equipment

Wire-mesh wheel

Moon Buggy
This open car was taken to the Moon for the first time in *Apollo 15*. Its correct name is the lunar roving vehicle.

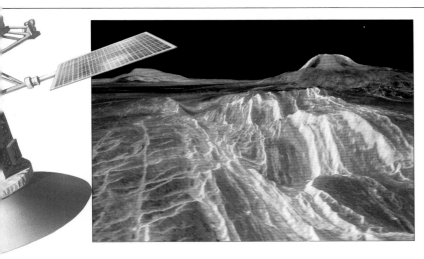

Venera 9 Venus landing

The space probe was in a capsule on the Venera *spacecraft.*

The capsule fell through the atmosphere of Venus.

The heat-shield covers separated and fell off.

Wish You Were Here?

The *Magellan* probe used radar cameras to take pictures through the thick fog around Venus. Computers made this 3-D image of the volcanoes.

The brake is shaped like a disk to help slow the space probe down.

The probe was slowed down by a small parachute.

Venus *Venera*

Several Russian missions have been to Venus. This program was called Venera. The spacecraft sent pictures back to Earth. This is the part of *Venera 9* that went down to Venus by parachute.

Instrument container

The landing ring helped make the landing soft.

Three larger parachutes were used for the final stage.

VENUS

Hot Orange

Venus is the hottest planet of all – so hot that it could melt lead! It has a bright orange sky with flashes of lightning. The Earth spins around once every 24 hours, but Venus spins very slowly – once every 244 Earth days!

After a safe landing, the television cameras and instruments were switched on.

THE RED PLANET

Viking spacecraft

The Viking lander is folded into a capsule on the spacecraft.

It leaves the orbiter and begins its journey down to Mars.

Mars is called the red planet because the soil and rocks are red. Light winds blow dust around, which makes the sky look pink. People once thought there was life on Mars, but nothing living has been found so far.

The *Viking* spacecraft were sent to Mars to find out what it is like. Two missions, *Viking 1* and *Viking 2*, made the journey. Perhaps one day people may go to live on Mars because it is the planet most like our own.

The television camera takes a series of pictures as it moves around.

It moves so fast that it gets very hot.

A parachute is used to slow it down. Then the heat shield drops off.

This remote-control arm is used to collect samples of Martian soil.

Tight Fit
The *Viking* lander fits into a capsule on the spacecraft. With its legs folded up, it looks a bit like a tortoise inside its shell.

The legs unfold, and the rockets are used as brakes for a soft landing.

The *Viking* lander

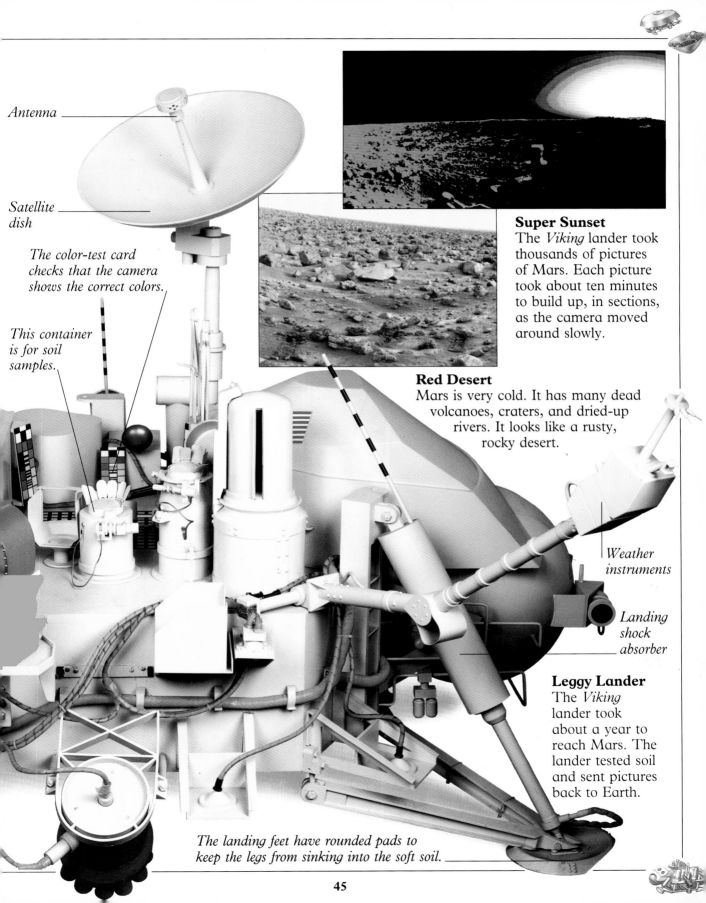

Antenna

Satellite dish

The color-test card checks that the camera shows the correct colors.

This container is for soil samples.

Super Sunset
The *Viking* lander took thousands of pictures of Mars. Each picture took about ten minutes to build up, in sections, as the camera moved around slowly.

Red Desert
Mars is very cold. It has many dead volcanoes, craters, and dried-up rivers. It looks like a rusty, rocky desert.

Weather instruments

Landing shock absorber

Leggy Lander
The *Viking* lander took about a year to reach Mars. The lander tested soil and sent pictures back to Earth.

The landing feet have rounded pads to keep the legs from sinking into the soft soil.

JUPITER AND SATURN

These two giants are the largest planets in our solar system. Jupiter is made of liquid, so it is not solid enough to land on, but if you could drive a car around its equator, it would take you six months of nonstop traveling. A similar journey around the Earth's equator would take only two weeks. Saturn is a beautiful planet with shining rings around its middle. Both planets spin around very fast, pulling the clouds into stripes.

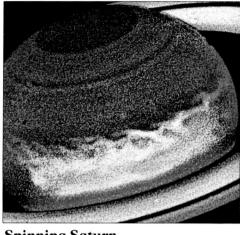

Spinning Saturn
Saturn is a giant, spinning ball of liquid held together by gravity. This photograph shows a band of clouds and the rings.

S A T U R N

A power supply is carried on the probe. It does not use solar power because it is working so far from the Sun.

Radio antenna

This disk has pictures of the Earth and sounds, such as a baby crying and music. If aliens find the disk, it will tell them about Earth.

Dish antenna

Television cameras

Seven Cold Rings
Saturn's rings are made up of glittering pieces of ice like trillions of snowballs.

***Voyager* Voyages**
The Voyager project sent probes to explore Jupiter and Saturn. *Voyager 1* did its job so well that *Voyager 2* was rerouted to go to Uranus and Neptune.

Pioneer's **Pictures**

The program for sending unmanned spacecraft to Jupiter was called Pioneer. *Pioneer 10* succeeded, so *11* went on to Saturn. Both sent back lots of pictures.

Power supply

Asteroid and meteor detector

Pioneer

Dish antenna

Mega Moons

Jupiter has 16 moons circling around it. The largest is called Ganymede. It is bigger than Mercury.

Sun sensor

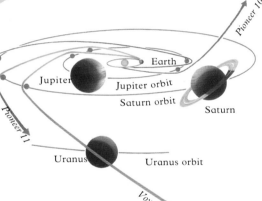

Pioneer 10

Earth

Jupiter

Jupiter orbit

Saturn orbit

Saturn

Pioneer 11

Uranus

Uranus orbit

Voyager 2

Neptune

One-way Ticket

The journeys of *Pioneer 10* and *11* and of *Voyager 1* and *2* passed several of the planets. These spacecraft are now heading for the stars.

J U P I T E R

Red Storm

Jupiter, like Saturn, is a huge ball of liquid. It has icy clouds and a giant red spot that is the center of a huge storm.

Swirling winds blow Jupiter's clouds into a hurricanelike storm.

THE OUTER PLANETS

Uranus, Neptune, and Pluto are the farthest planets from the Sun, so they are called the outer planets. They are very cold. Uranus was the first planet to be discovered using a telescope, because you cannot see it from Earth just with your eyes. Pluto is the farthest away, and no spacecraft has visited it yet. If a jet could fly there, it would take a thousand years!

The rings of Uranus are made of rocks. The widest ring, called Epsilon, is 58 miles (93 km) across.

Sideways Spinner
Uranus looks as if it is spinning on its side. It is covered in dense fog.

Outer solar system

Orbit of Pluto

Ellipses
The planets move around the Sun in squashed circles called ellipses. This girl is drawing ellipses.

Order of Orbits
"Planet" means "wanderer." The planets travel around the Sun in paths, called orbits. The ones nearer the Sun have shorter lengths of orbit than the ones farther away. Some astronomers believe there is a tenth planet, not yet discovered. They call it Planet X and it may be bigger than Pluto.

This diagram shows the orbits of the planets and their places in the solar system. It does not show the correct sizes of the planets.

NEPTUNE

Blue Neptune
The *Voyager 2*
spacecraft took
photographs of
Neptune's clouds
and its storm, called
the Great Dark Spot.

Inner solar
system

PLUTO

Brrrrr!
Pluto is tiny – smaller
than our Moon – and
at night is ten times
colder than a freezer!

*Light from the Sun takes
three minutes to reach
Mercury and eight
minutes to reach the
Earth, but five and a half
hours to reach Pluto!*

The Sun
Orbit of Mercury
Orbit of Venus
Orbit of Earth
Orbit of Mars

*Pluto is not always
the farthest planet
from the Sun.
Its strange orbit
means that
sometimes it is
nearer than
Neptune.*

Orbit of Jupiter
Orbit of Saturn
Orbit of Uranus
Orbit of Neptune

SKY WANDERERS

Path of Halley's comet

Between Mars and Jupiter, there is a belt of rocks in orbit, called the asteroid belt. The chunks of rock are called asteroids. Sometimes these rocks crash into each other and bits fall down toward Earth.

Also in orbit around the Sun are lumps of rock and ice, called comets. When comets get near the Sun, they shine like "hairy stars." This is what people used to call them long ago. The most famous comet is Halley's comet, named after the man who first studied it.

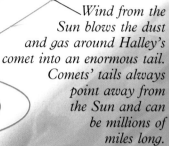

Wind from the Sun blows the dust and gas around Halley's comet into an enormous tail. Comets' tails always point away from the Sun and can be millions of miles long.

Solar System

Halley's comet

Earth Comets
Because they are made of rock and ice, comets are often called "dirty snowballs." Make your own comet the next time it snows!

Regular Visitor
We see Halley's comet from Earth once every 76 years because it takes that long to orbit the Sun. It was shown in a picture called the Bayeux Tapestry more than 900 years ago!

The comet's center is made of ice. As it gets near the Sun, the ice melts.

A Belt You Cannot Wear
The asteroids in the asteroid belt are really mini-planets. There are thousands of them. The largest is about 600 miles (1,000 km) across.

Comet Head
This is a photograph of the head of Halley's comet. Computer colors show the bright center and the layers around it.

Meteor Shower

If a lump of rock or metal burns up before it reaches the ground, it is called a meteor or shooting star. This photograph shows lots of them falling together in a meteor shower.

Crash! Bang!

A large meteor that does not burn up as it plunges through the Earth's atmosphere is called a meteorite.

It travels so fast, it shatters into pieces as it hits the ground.

It causes shock waves as it lands.

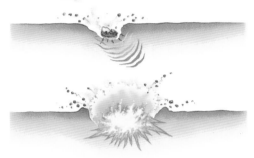

The explosion leaves a big hole, called a crater.

This huge meteorite crater is in Arizona.

Dish antenna

Solar cells for power

Giotto

Camera

Dust shield

Comet Quest

In 1986, *Giotto*, a European spacecraft, passed very close to Halley's comet and took pictures of it. *Giotto* is now searching for other comets.

Gaspra the Asteroid

No one had seen a picture of Gaspra until the U.S. spacecraft *Galileo* took this one in 1991 as it passed the asteroid belt.

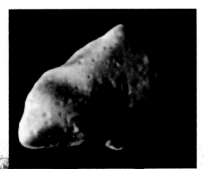

SKY WATCHING

If you look up at the sky on a clear night, you can see hundreds of stars and, often, the Moon. But if you use binoculars or a telescope, you can see even more – for example, the craters on the Moon and the planets.

When astronomers study the universe, they use huge radio telescopes, some with dishes, to help them see far, far away, and to gather information from space. The Hubble Space Telescope is the largest telescope to be put into space. It can provide clear pictures of stars and galaxies because it orbits 380 miles (612 km) above the Earth's murky atmosphere.

Clearly Venus
This photograph of Venus was taken by the *Pioneer* Venus orbiter. It used radar to get a clear picture through the thick clouds around Venus. The signals were sent back to Earth to a radio telescope, where this picture was produced.

Solar panel

Radio telescope

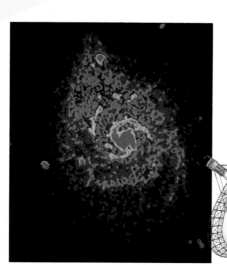

Whirligig
This radio map of the Whirlpool galaxy was taken by a radio telescope. The added colors show the spiral arms of the galaxy.

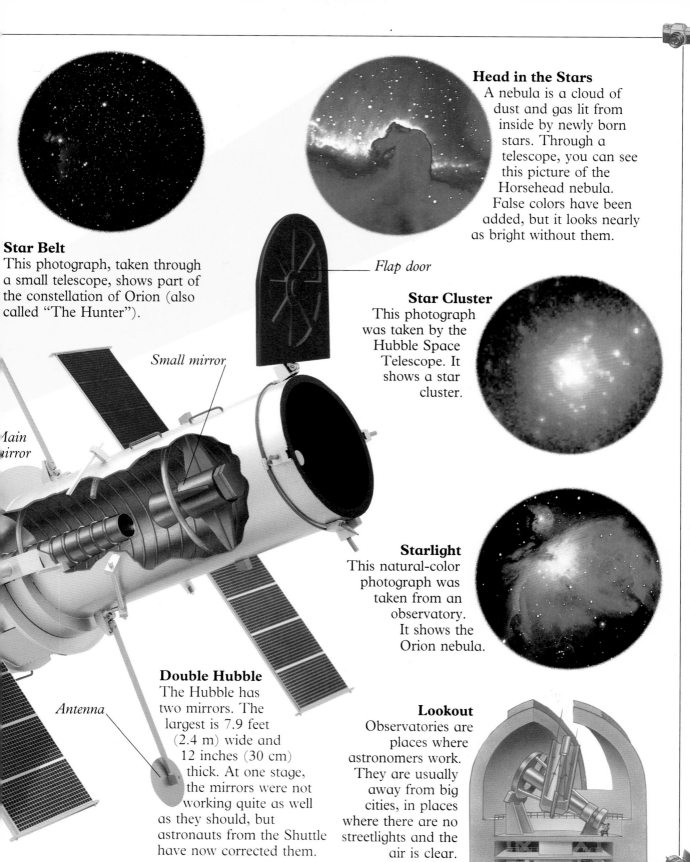

Head in the Stars

A nebula is a cloud of dust and gas lit from inside by newly born stars. Through a telescope, you can see this picture of the Horsehead nebula. False colors have been added, but it looks nearly as bright without them.

Star Belt

This photograph, taken through a small telescope, shows part of the constellation of Orion (also called "The Hunter").

Flap door

Small mirror

Main mirror

Star Cluster

This photograph was taken by the Hubble Space Telescope. It shows a star cluster.

Starlight

This natural-color photograph was taken from an observatory. It shows the Orion nebula.

Double Hubble

The Hubble has two mirrors. The largest is 7.9 feet (2.4 m) wide and 12 inches (30 cm) thick. At one stage, the mirrors were not working quite as well as they should, but astronauts from the Shuttle have now corrected them.

Antenna

Lookout

Observatories are places where astronomers work. They are usually away from big cities, in places where there are no streetlights and the air is clear.

STARS AND GALAXIES

Stars look like tiny points of light from the Earth, but really they are huge, hot balls of burning gas deep in space. They are forming, changing, and dying all the time. There are big stars, called giants, even bigger ones, called supergiants, and small ones, called dwarfs. Our Sun is just one of about one hundred billion stars that all belong to a galaxy called the Milky Way. A galaxy is a group of millions of stars, held together by a strong force, called gravity.

Starry, Starry Night
On a clear night, do not forget to look up at the sky! You will see hundreds of twinkling stars, like tiny sparkling diamonds, far above you.

The gas and dust pack tightly together, getting smaller and very hot.

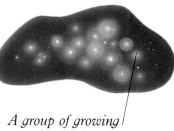

A group of growing stars is called a cluster.

A new star is very bright. It shines steadily for many years.

As it cools, the star gets bigger and forms a red giant.

A star is born inside a huge cloud of dust and gas, called a nebula. The word "nebula" means "mist."

Near the end of its life, the core of a red giant may cave in and give off layers of gas.

Sky Lights
There are many new stars in the gas and dust of this pink nebula. A new young star on its own shines blue.

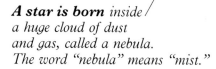

Sometimes a giant star explodes and is blown to pieces. This is called a supernova.

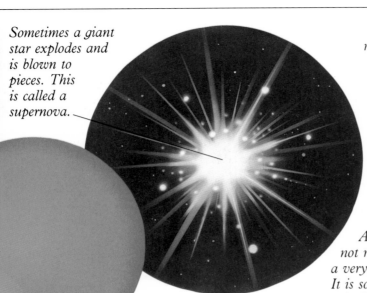

A supernova explosion sometimes results in a pulsar. A pulsar is a rapidly spinning star that gives off pulses of radio waves.

A black hole is not really a hole, but a very tightly packed object. It is solid and does not reflect any light, so it looks like a hole! Its gravity pulls things toward it like water down a drain.

Some dying stars grow into huge red supergiants.

Our solar system

Massive stars shine very brightly but do not live as long as smaller stars.

If seen through a telescope, the star now looks like a planet, so it is called a planetary nebula.

Some stars gradually get smaller and whiter until they become white dwarf stars.

Star Spinner

Our galaxy, called the Milky Way, is a spiral galaxy. Our solar system is about two-thirds of the way out from the center, in one of the spiral arms. There are lots of galaxies in the universe, and they have different shapes. Try painting some!

Galaxies

A spiral galaxy

An elliptical galaxy

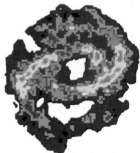

A barred spiral galaxy

CHAPTER 2

THE NATURAL WORLD

When Earth was brand-new more than 4.5 billion years ago, there were violent storms, lightning bolts, and fiery volcanoes on its surface, but there was no life. In time, warm, shallow seas formed, and it was here that the first life appeared in the form of simple plants, called bacteria and algae. Soon, more complex plants developed, and, by about 600 million years ago, tiny, soft animals were living in the water, too. Later, these became larger and developed spines, and small plants began to grow on the shores.

Insects were the first animals on land, followed by amphibians (which live in water and on land), then dinosaurs, birds, and mammals. It was not until quite recently – about 5 million years ago – that humans first appeared.

Plant Life
Sea Life
Insects and Spiders
Dinosaurs
Birds
Mammals
The Human Body

PLANT LIFE

Plants grow everywhere, from the icy Arctic to the tropics – anywhere there is air, light, and water.

In deserts, where it rarely rains, plants have to save moisture. Some, such as cacti, have pleated stems that expand to store water when there is a shower. Where the climate is hot and humid, plants grow very quickly all year round to make lush rain forests. In temperate places – where it is not too hot and not too cold, with a medium amount of rain – most plants flower and fruit in summer, but lose their leaves in winter. Plants usually grow in soil, but some absorb what they need directly from the water.

Without plants, our world would be very different. From plants, we get food, clothes, paper, and lots of other things. Most important of all, plants provide oxygen, which all living creatures need to breathe.

Water hyacinth

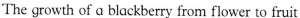

The growth of a blackberry from flower to fruit

Stag's-horn sumac

Cobra
lily

Larch cones

Tropical rain forest

Acorns

Oak leaf

Oak tree in winter and in summer

WHAT IS A PLANT?

The leaves are used for making food.

An apple tree and a cactus do not look much like each other. But they have more in common than you might think. They are both plants. Like all other plants, they make their own food, and during the course of their lives, they can produce many new plants. To do these things, they use their roots, stems, leaves, and flowers. You will find that no matter how different one plant may look from another, each one is using these parts in much the same way in order to live and grow.

Look, No Hands
Some plants do not need soil. They prefer to perch in trees. A stag's horn fern gets a good grip by wrapping its large fronds around a branch.

Trees have a main stem and many branches.

Apples contain seeds.

Hitching a Lift
Some plants use other plants for support. This passionflower can travel long distances by twining itself around the trunks and branches of trees.

The flowers make seeds for new plants.

The flexible stems can bend and twine.

Fresh Fruit
What do apple trees grow? Apples, of course! But they also grow flowers – and without them, there would never be any apples.

Good Fronds

Fern leaves are called fronds, and they grow straight out of the ground. Their stems are beneath the soil.

Feather Duster

Pampas grass has long, sharp leaves, and those feathery plumes are its flowers.

This flower bud will open into a flower.

Don't Touch!

This cactus has a stem with branches, but where are its leaves? The prickles are really special leaves.

Eye-catching

You can't miss these flowers. Like other plants that flower, orchids want to make quite sure that insects come visiting. So they advertise themselves with colors and scents that are attractive to insects.

Nonflowering Plants

Most plants flower, but not all. Ferns, mosses, liverworts, and lichens don't. The earliest plants to colonize the earth – over 300 million years ago – were non-flowering. These plants are their descendants. Instead of seeds, they produce tiny spores, which become new plants.

Liverwort

Lichen

Moss

The curling tendrils hang on tightly.

LEAVES

Leaves work very hard for plants. They make food, and they also help plants cope with serious problems, such as how to survive the cold or get enough water. Leaves come in all sorts of shapes and sizes – large and small, thick and thin. In fact, you can tell quite a lot about a plant and where it grows just by looking at its leaves.

Weatherproof

Scotch pine trees need to be tough to survive long, cold winters. They have thousands of tiny, needlelike leaves. The needles have a waterproof coating to protect them from rain and snow.

Wind blows through the needles without damaging them.

Water Store

Agave plants grow in hot places where it may not rain for weeks on end. They are able to store water in their large, thick leaves.

Agave leaves can grow up to seven feet (two meters) long.

Open Sesame

Leaves have tiny holes, called stomata, which the plant can open and close. When the stomata are open, they let air in and out, and water out. When they are closed, water can't escape from the leaves.

Shapes and Sizes

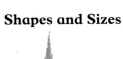

Japanese maple   Himalayan birch   Fig   Acacia   Horse chestnut

Drip-dry

Life in a tropical forest is hot and damp, and there is no shortage of water. Monstera leaves have a special waxy surface, so the water can run off.

Monstera plants grow in the shade of trees that constantly drip moisture.

Prickly Customers

Plants can't run away from hungry animals, so they have to protect themselves. Prickly holly leaves are left alone!

Tough, glossy leaves are a good defense against wind and weather.

Shady Character

The maidenhair fern lives in damp, shady places, where its fragile leaves won't dry out in the sunlight.

Water Signals

Leaves can't talk, but they can sometimes send a message. The leaves of this cyclamen are limp and drooping, as if the plant were unhappy. The soil in the pot is dry, and the message is, "Water me."

Do Plants Sweat?

Plants are constantly losing water through their leaves as part of a process called transpiration. Most of the time you can't see it happening. But if you put a plant inside a plastic bag and fasten it, after a while you will see water drops on the inside of the bag. The moisture you can see is coming from the leaves of the plant.

HOW PLANTS MAKE FOOD

Hungry animals can go out hunting for their food – but plants cannot. Instead, they make their own food in their leaves, using light from the Sun, water from the soil, and carbon dioxide from the air. A plant's way of making food is called photosynthesis. It takes place during the day when the leaves are absorbing sunlight.

Reach for the Sun
These palms grow in the shade of tall trees. Their leaves are arranged like fans to help them catch all the light they can.

The leaves of all plants contain a special pigment, which gives them their green color. It is called chlorophyll.

Colorful Cover-up
All leaves contain green chlorophyll. But in some leaves, the green is hidden from sight by other, stronger colors.

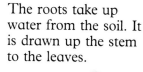

The roots take up water from the soil. It is drawn up the stem to the leaves.

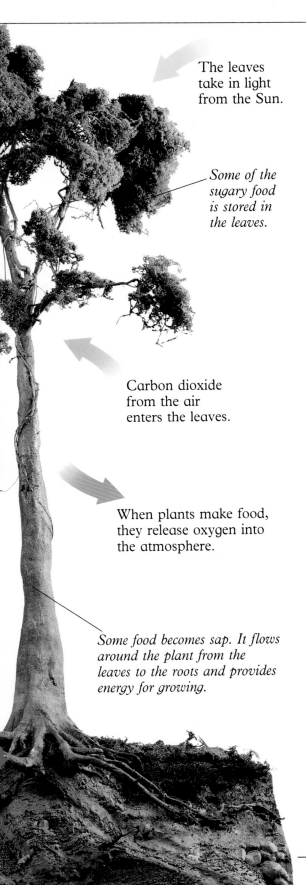

The leaves take in light from the Sun.

Some of the sugary food is stored in the leaves.

Carbon dioxide from the air enters the leaves.

When plants make food, they release oxygen into the atmosphere.

Some food becomes sap. It flows around the plant from the leaves to the roots and provides energy for growing.

Rest Time
Without sunlight, plants cannot make food. When it is dark, they shut down for the night by closing their stomata.

Photosynthesis
Chlorophyll in the leaves absorbs sunlight. Sunlight provides the plant with energy to turn water and carbon dioxide into food.

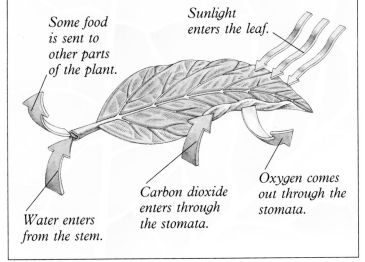

Some food is sent to other parts of the plant.

Sunlight enters the leaf.

Water enters from the stem.

Carbon dioxide enters through the stomata.

Oxygen comes out through the stomata.

Bare tree in winter

Tree in summer leaf

Hibernation
In winter, there is less light, and the water often freezes in the ground. It is difficult to make food, so plants grow very little at this time of year. Many trees shed their leaves.

CARNIVOROUS PLANTS

Water Jugs
A pitcher plant has several pitchers, so it can catch a lot of flies.

Watch out! These plants are meat-eaters, and they have some very cunning devices for trapping their victims. Unlucky insects, attracted by the plant's scent and color, discover too late that they have been tricked. It is a nasty end for an insect, but a ready-made, nutritious meal for the plant.

The lid can close to keep rainwater out.

Gruesome Gruel
Flies lose their footing on the slippery rim of the hanging pitcher plant and tumble into the water below. They gradually dissolve into a kind of fly broth.

Rim with nectar

Pitcher for collecting water

Remains of flies

Venus's-flytrap
The instant an unsuspecting insect lands, the Venus's-flytrap snaps into action.

To a fly, this pad looks like a safe landing place.

Swamped
Most carnivorous plants, like these cobra lilies, grow in boggy places where the minerals they need are in short supply. The insects they catch make a vital addition to their diet because they are rich in the missing minerals.

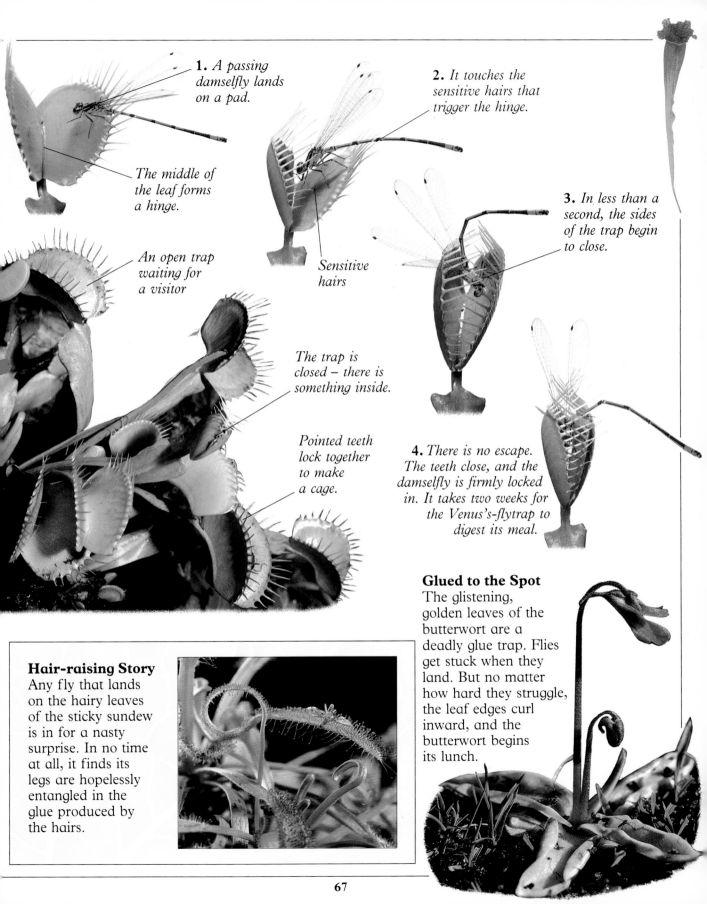

1. *A passing damselfly lands on a pad.*

The middle of the leaf forms a hinge.

An open trap waiting for a visitor

2. *It touches the sensitive hairs that trigger the hinge.*

Sensitive hairs

3. *In less than a second, the sides of the trap begin to close.*

The trap is closed – there is something inside.

Pointed teeth lock together to make a cage.

4. *There is no escape. The teeth close, and the damselfly is firmly locked in. It takes two weeks for the Venus's-flytrap to digest its meal.*

Hair-raising Story

Any fly that lands on the hairy leaves of the sticky sundew is in for a nasty surprise. In no time at all, it finds its legs are hopelessly entangled in the glue produced by the hairs.

Glued to the Spot

The glistening, golden leaves of the butterwort are a deadly glue trap. Flies get stuck when they land. But no matter how hard they struggle, the leaf edges curl inward, and the butterwort begins its lunch.

STEMS

Roots, shoots, leaves, and flowers are all connected to the plant's stem. Although it may not always stand as straight, the stem is rather like your backbone, holding all the different parts together. Being in the center of everything means it is in the perfect position to carry water and food to every part of the plant.

Clever Creature
This aphid knows just where to go for food supplies! It takes less than a second to pierce the soft part of the stem, which is full of nutritious sap.

This young tree is two years old.

Sun Worshipers
The stems of sunflowers turn so their flowers can always face the sun.

Branches grow out from the stem and hold the leaves out to the light.

As the tree gets older and taller, the main stem will thicken to form a trunk.

Clinging On
In its rush to the light, the sweet pea has no time to grow strong stems. Instead, it uses twirling tendrils to wrap around other plants. They will support its fast climb to the top.

The Widest Spread on Earth
Just one banyan tree can make a forest! Its branches throw down special aerial roots. These grow into the ground and expand into trunks. A single tree in Calcutta has over 1,000 of these trunk look-alikes.

Water and minerals travel up the stem from the roots.

Supporting Role

The strong, hard stems of bamboo are called canes. In some parts of the world, they make huge thickets 23 feet (7 meters) high.

Weak at the Knees

Gourd stems don't even try to stand up to support their fruit, which may weigh several pounds. They just trail gracefully over the ground.

The branches are slender and not yet very strong.

The inside of a bamboo cane is hollow.

Food is made in the leaves and travels down to the roots.

Keep Off !

It is quite difficult to get near a thistle, but any animal that does won't be back for a second bite.

Close-up of thistle stem

Section through cactus

Water Diet

African baobab trees have a special way of surviving. During the rainy season, their trunks store so much water that they visibly swell up. These reserves get them through the hot, dry times, when their slimmer shape slowly returns.

Conservation

Desert cacti hold water reserves in their thick stems. Thirsty animals know that. But the barricade of fierce spines means no free drinks.

ROOTS

Roots are not pretty or colorful like leaves and flowers, but plants couldn't do without them! Anchored in the soil, they hold plants upright against wind and weather. They also grow out and down in search of water and minerals, which are drawn all the way up to the leaves. Think how tall a tree can grow, and you will see it needs strong roots to keep it supported.

There are little pockets of air in the soil. Without air, roots would wither and die.

Roots can fit themselves into tiny spaces.

When earthworms burrow, they help to add air to the soil.

The roots of the tree grow outward to balance the spread of the branches above.

Knobbly Knees
Swamp cypresses grow in boggy ground, where there is not enough air. These strange bumps probably supply the roots with the air they need. They are known as cypress knees.

Strong Intent
Roots don't let much stand in their way. These roots are growing toward the drain in the road, where there is a useful supply of water.

Rootless Wonder
Draped like strange beards over the branches of trees, the extraordinary Spanish moss plant survives with no roots at all. Spanish moss grows in subtropical climates where the air is very wet. It absorbs all the moisture it needs through its fine, threadlike leaves.

Rock Climbers
Alpine plants grow against rock faces to protect themselves from high winds and icy squalls. Their tiny roots wriggle into cracks in the rock.

Most roots grow in the top 12 inches (30 centimeters) of soil. This part contains most of the important minerals the tree needs.

Every root grows a mass of tiny hairs near its tip to absorb water from the soil.

Water Crops
Plants need water, minerals, and some support for their roots. But they do not necessarily need soil. Today, many food crops are grown entirely in water filled with special pebbles. They are given liquid minerals to replace those in the soil.

INSIDE A FLOWER

Hibiscus

When you look at flowers, you notice many colors, shapes, and sizes. Some plants have a single flower. Others have so many, it is impossible to keep count. But stop and take a closer look – this time, inside. However different they may look, flowers all have the same basic parts. This is because all plants produce flowers for the same purpose: to make seeds so another plant can grow.

New Beginnings
The process of making new seeds is called reproduction. The male and female parts in the center of this lily are its reproductive parts.

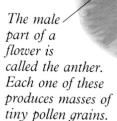

The male part of a flower is called the anther. Each one of these produces masses of tiny pollen grains.

Insects leave pollen on the female part, which is called the stigma.

Lily

Mistaken Identity
You could be confused by this poinsettia. What look like bright red petals are actually a kind of leaf, called a bract. The real flowers are the tiny green dots in the center.

Grand Finale
Not all plants flower every year, but there is no other plant which is as slow as the *Puya raimondii* from South America. It takes 150 years to produce a massive flower spike, up to 30 feet (10 meters) tall. Exhausted, it then dies, but luckily, not before it has produced a few seeds.

Mighty Magnolias

The last of the dinosaurs may have munched on magnolias like these. Magnolia trees are one of the oldest flowering plants. They have been around for 100 million years, and they are still growing today.

The petals are brightly colored, with special markings to attract insects.

Flower Arrangement

All these flowers have the same basic parts, but they are arranged on the stem in different ways.

These bell-shaped flowers hang down.

Is This a Flower?

Tropical orchids like this one often look more like strange insects than flowers.

Hundreds of small flowers grow in a single spire.

The flowers of the spider orchid can be up to 2 feet (60 centimeters) long.

Daisy petals are arranged like the rays of the Sun.

Each of these tightly packed flowers is called a floret.

Poppy petals open out to the light.

Each tiny point is a flower.

Snake's head fritillary

Allium

Mullein

Transvaal daisy

Poppy

Yarrow

FLOWERS AND THEIR POLLINATORS

Most plants cannot make seeds without some outside help. The first job is to move pollen from the anther of one flower to the stigma of another. This is called pollination. Plants cannot travel, but their flowers produce sweet nectar, which animals love. As the animal feeds on the nectar, some pollen rubs onto its body. Each time it moves on to another flower, it leaves some pollen behind and picks up a new supply.

Honey Hunters

As it feasts on nectar, the Australian honey possum gets pollen on its fur.

Pollen Stop

Bees flit from flower to flower all day, feeding on nectar. Each time they stop, they pick up some pollen.

This bee is having a good pollen bath.

A flower, not a fly!

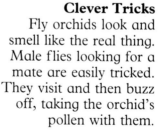

Clever Tricks

Fly orchids look and smell like the real thing. Male flies looking for a mate are easily tricked. They visit and then buzz off, taking the orchid's pollen with them.

Pollen from the anthers of the flower sticks to the butterfly's body.

Nectar Gatherers

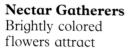

Brightly colored flowers attract butterflies looking for nectar.

Pollen Galore

Some flowers are either male or female. A catkin, for example, is its plant's male flower. It produces masses of pollen to make sure some will be blown onto the stigmas of the tiny female flowers.

Special Collection

When the hummingbird pushes its long beak deep inside the flower to collect the nectar, some pollen brushes off onto its body.

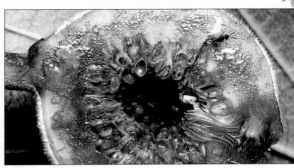

Inside Story

Not many insects would ever find the flowers of the fig tree. They actually grow inside the figs! They are pollinated by special fig wasps that live inside the fig. When the flowers are producing pollen, some of the wasps leave home. They move into another fig, carrying pollen with them on their bodies.

Pollen is brushed onto the bat's fur as it moves from flower to flower.

The bat's long tongue is perfect for whisking out the nectar.

Tropical Favorite

The bird-of-paradise flower grows in the tropics. It is pollinated by bats as well as by birds.

On its nightly nectar hunt, one bat can pollinate several flowers.

Bats find it easy to pick out the spiky shape of the bird-of-paradise flower at night.

FLOWERS BECOME FRUITS

After they have been pollinated, flowers produce seeds and fruits. The fruits protect the seeds and keep them safe until the time comes for them to grow. Like flowers, fruits come in all sorts of shapes and sizes. Pine trees make cones, dandelions make parachutes, and plum trees make plums. Every fruit has its own kind of seed. Some are light enough to be blown away on the wind. Others are armor-plated so they can be swallowed by animals and pass out in their droppings without being damaged!

From Rose to Hip
Bees are attracted to this rose by its sweet smell and the promise of nectar.

1. *The swelling beneath the flower is called the receptacle, and it will become the fruit.*

Stigma

Pollen grain tube

Petal

Ovary

Egg

Fertilization
An insect leaves some pollen on the stigma of a flower. Now fertilization can take place, and each tiny pollen grain grows a long tube. The tube grows down until it reaches the ovary, where eggs are produced. Now a male gamete from the pollen tube joins with an egg from the ovary, and a seed is born.

Leaving Home
Seeds need space and light to grow. If they fall straight off the parent plant, they have to struggle to grow in its shadow. So plants use all sorts of clever devices for making sure their seeds are carried away from them by wind or animals. Some have exploding pods that catapult the seeds into the air.

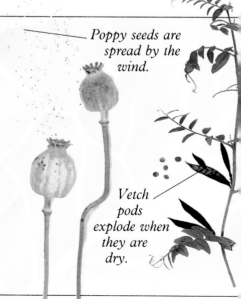

Poppy seeds are spread by the wind.

Vetch pods explode when they are dry.

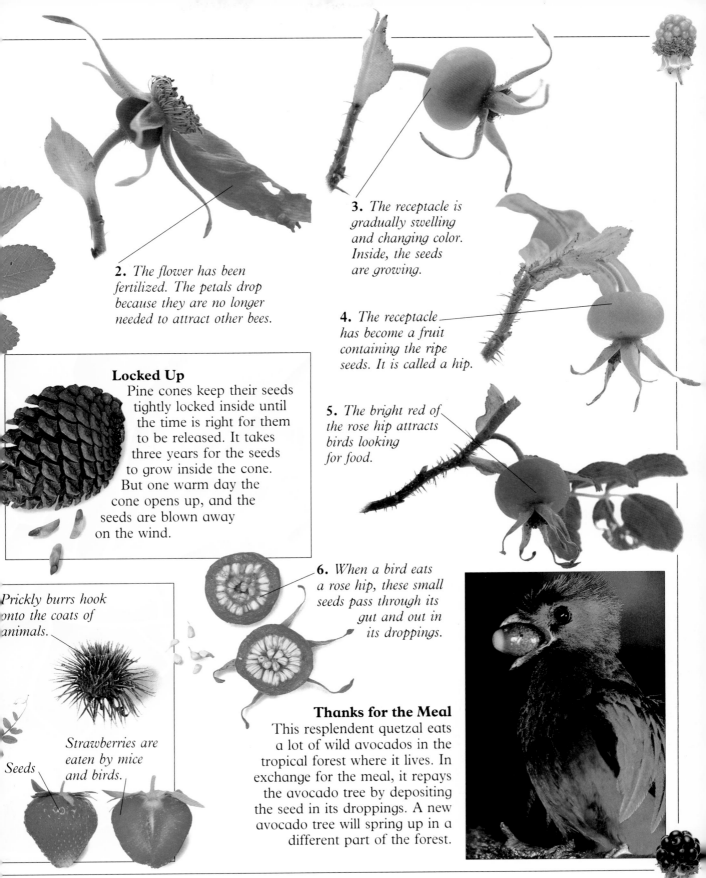

2. *The flower has been fertilized. The petals drop because they are no longer needed to attract other bees.*

3. *The receptacle is gradually swelling and changing color. Inside, the seeds are growing.*

4. *The receptacle has become a fruit containing the ripe seeds. It is called a hip.*

5. *The bright red of the rose hip attracts birds looking for food.*

Locked Up

Pine cones keep their seeds tightly locked inside until the time is right for them to be released. It takes three years for the seeds to grow inside the cone. But one warm day the cone opens up, and the seeds are blown away on the wind.

6. *When a bird eats a rose hip, these small seeds pass through its gut and out in its droppings.*

Prickly burrs hook onto the coats of animals.

Strawberries are eaten by mice and birds.

Seeds

Thanks for the Meal

This resplendent quetzal eats a lot of wild avocados in the tropical forest where it lives. In exchange for the meal, it repays the avocado tree by depositing the seed in its droppings. A new avocado tree will spring up in a different part of the forest.

SEEDS BECOME PLANTS

Below ground, a seed is waiting to start life. But until it gets the right signals, a seed will remain just a seed. As soon as the soil becomes warm and damp, the seed can begin to absorb moisture. This makes it swell, and the seed case splits open. Germination has begun – and the seedling starts to grow toward the light.

Wallflowers
Some seeds land in odd places – and there they grow!

The leaves unfold into a fan shape.

Peanuts

Peanut bush

Coffee bush

Seed and Plant
You may recognize these seeds, but do you know what they grow into?

Coffee beans

Oak tree

Acorns

Lemon pips

Lemon tree

Roots grow down through the husk of the coconut.

Fast Food
Mustard and cress seeds are quick and easy to grow. Put some damp cotton in clean eggshells and sprinkle a few seeds on top. Look at the seeds each day and keep the cotton damp. In about ten days, you will have your own homegrown salad!

Mustard and cress seeds

Damp cotton

On the Beach
A coconut can travel long distances by sea, until it is washed up onto the shore. In the warm sand, it sprouts and starts to grow.

Some Like It Hot
Some seeds must literally go through fire and water before they can germinate. In the arid Australian outback, fierce fires can rip across the land. They wake the dormant seeds of acacia trees.

As the shoot gets longer and thicker, the first leaves open.

The stem is formed from the stalks of the leaves.

The coconut contains liquid, so the seedling has its own water supply for a while.

Inside the bean, you can see the part that is the future root.

A first sign of life – the root pushes through the skin of the bean.

The shoot appears above ground. This is the stem.

Roots and Shoots
When the soil gets warm in spring, this fava bean germinates and begins to grow.

GROWING WITHOUT SEEDS

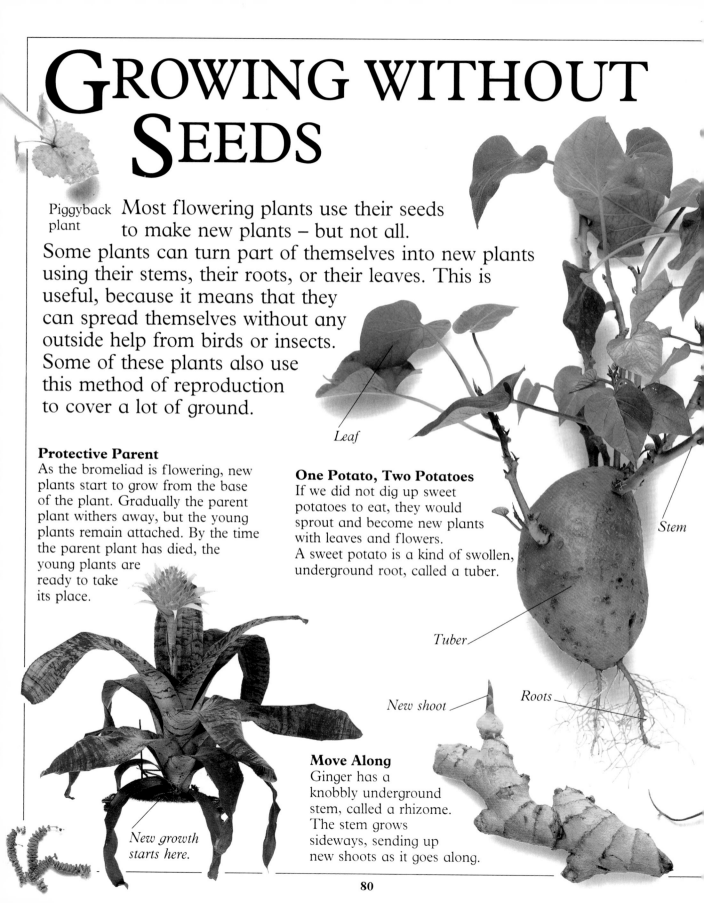

Piggyback plant

Most flowering plants use their seeds to make new plants – but not all.
Some plants can turn part of themselves into new plants using their stems, their roots, or their leaves. This is useful, because it means that they can spread themselves without any outside help from birds or insects. Some of these plants also use this method of reproduction to cover a lot of ground.

Leaf

Protective Parent
As the bromeliad is flowering, new plants start to grow from the base of the plant. Gradually the parent plant withers away, but the young plants remain attached. By the time the parent plant has died, the young plants are ready to take its place.

One Potato, Two Potatoes
If we did not dig up sweet potatoes to eat, they would sprout and become new plants with leaves and flowers.
A sweet potato is a kind of swollen, underground root, called a tuber.

Stem

Tuber

New shoot

Roots

New growth starts here.

Move Along
Ginger has a knobbly underground stem, called a rhizome. The stem grows sideways, sending up new shoots as it goes along.

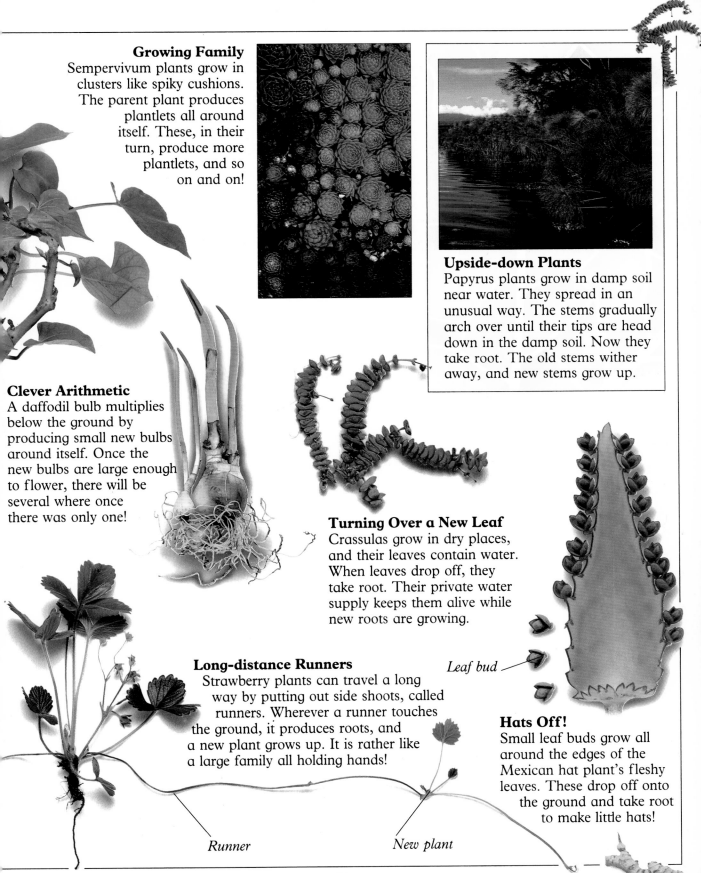

Growing Family

Sempervivum plants grow in clusters like spiky cushions. The parent plant produces plantlets all around itself. These, in their turn, produce more plantlets, and so on and on!

Upside-down Plants

Papyrus plants grow in damp soil near water. They spread in an unusual way. The stems gradually arch over until their tips are head down in the damp soil. Now they take root. The old stems wither away, and new stems grow up.

Clever Arithmetic

A daffodil bulb multiplies below the ground by producing small new bulbs around itself. Once the new bulbs are large enough to flower, there will be several where once there was only one!

Turning Over a New Leaf

Crassulas grow in dry places, and their leaves contain water. When leaves drop off, they take root. Their private water supply keeps them alive while new roots are growing.

Leaf bud

Long-distance Runners

Strawberry plants can travel a long way by putting out side shoots, called runners. Wherever a runner touches the ground, it produces roots, and a new plant grows up. It is rather like a large family all holding hands!

Hats Off!

Small leaf buds grow all around the edges of the Mexican hat plant's fleshy leaves. These drop off onto the ground and take root to make little hats!

Runner

New plant

TREES

Monkey-puzzle tree

Trees live longer than almost anything else on earth. They grow strong, woody trunks so that they can tower above other plants and get plenty of light. There are trees that lose all their leaves at once when the weather gets cold. They are known as deciduous trees. Others are evergreen and shed a few leaves at a time throughout the year.

The heartwood is the oldest part of the tree. It is no longer living, but it is very strong.

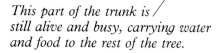

Bark is very important. It protects all the living, working parts of the trunk.

This part of the trunk is still alive and busy, carrying water and food to the rest of the tree.

Fiery Finale
Before they fall, the leaves of many deciduous trees change color. The green chlorophyll breaks down, revealing other colors that were hidden. Chemicals in the leaves may deepen these colors to fiery purples and reds.

Silver birch

Bark
As trees grow, their trunks expand. The bark cracks to make lots of different patterns.

Paperbark maple

Tibetan cherry

Holding Fast
Over the years, wind has shaped and battered the branches of this tree. But its trunk and strong roots keep it standing.

Deciduous Trees

Spanish oak

American mountain ash

Weeping willow

Cross Section

Every year a tree adds a new layer of growth to its trunk and branches. Look at this slice from a tree's trunk, and you will see lots of rings. Each ring shows the growth made by the tree in one year. By counting the rings, you can tell how old a tree is.

A wide ring shows that the tree grew a lot in this year.

A bad year for growth! Thin rings show how slowly trees can sometimes grow.

Shagbark hickory

Cork oak

Snake-bark maple

Evergreen Trees

Deodar

Mountain gum

Black spruce

The Incredible Hulk

What size waist do you have? The trunk of this giant sequoia in California measures 83 feet (25 m) around and stands 275 feet (83 m) high. It is so famous, it even has a name: General Sherman!

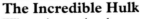

83

SEA LIFE

Almost three-quarters of the Earth is covered by oceans and seas. These billions of tons of salty water are home to silent sharks, playful dolphins, huge whales, spiny orange crabs, giant octopuses with their eight long arms, friendly turtles, and fish of all shapes and sizes. Beautiful underwater gardens of brightly colored coral provide a home for stinging sea anemones, poisonous sea slugs, and deadly eels.

Many parts of the world's seas and oceans are too deep, dark, and cold to support anything except small, simple forms of life. Other places are so fantastically deep that nobody has ever been able to explore them. One day, we may discover that even more incredible plants and animals than we imagined inhabit this silent, watery world.

Leatherback turtle

Coral reef

Blue whale

Tropical sundial shells

Flounder with
camouflage
coloring

Female anglerfish with males underneath

Edible crab

Harbor-seal pup

Common
starfish

PLANKTON

Seawater is full of billions of very tiny plants and animals called plankton. Most types of plankton are less than one millimeter long, but without them, very little else could live in the sea! They are the most important food for many fish, whales, and even birds. Their tiny size means that they are not strong enough to swim against the water's current, so they can only drift around.

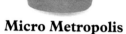

This curved feeler is covered in fine sensory hairs that help the copepod find food.

Animal plankton can be more than 100 times bigger than the plants they eat.

Plant plankton...eaten by...animal plankton...eaten by...fish...eaten by...seabirds

Micro Metropolis
Water fleas, called copepods, are a type of animal plankton. They are so small that more than one million could fit into a bucket of seawater.

Food for Everyone
Seabirds, sea turtles, and seals all eat other animals, such as fish and shellfish, that eat plankton. They would all starve if there were no plankton in the ocean.

The dark water shows through the colorless parts of this plant plankton.

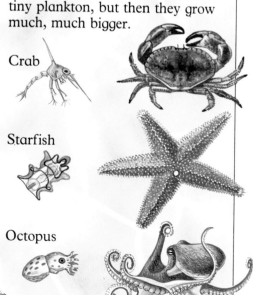

Part-time Plankton
Some animals begin their lives as tiny plankton, but then they grow much, much bigger.

Crab

Starfish

Octopus

Sun Lovers
Like all plants, plant plankton use sunlight to make their food.

There are many different kinds, or species, of copepods. This male is flashing bright blue to attract a female.

By waving these bristles, the copepod pushes water over its tiny gills and also traps plant plankton.

Countless Copepods
Copepods are the most common animals in the sea.

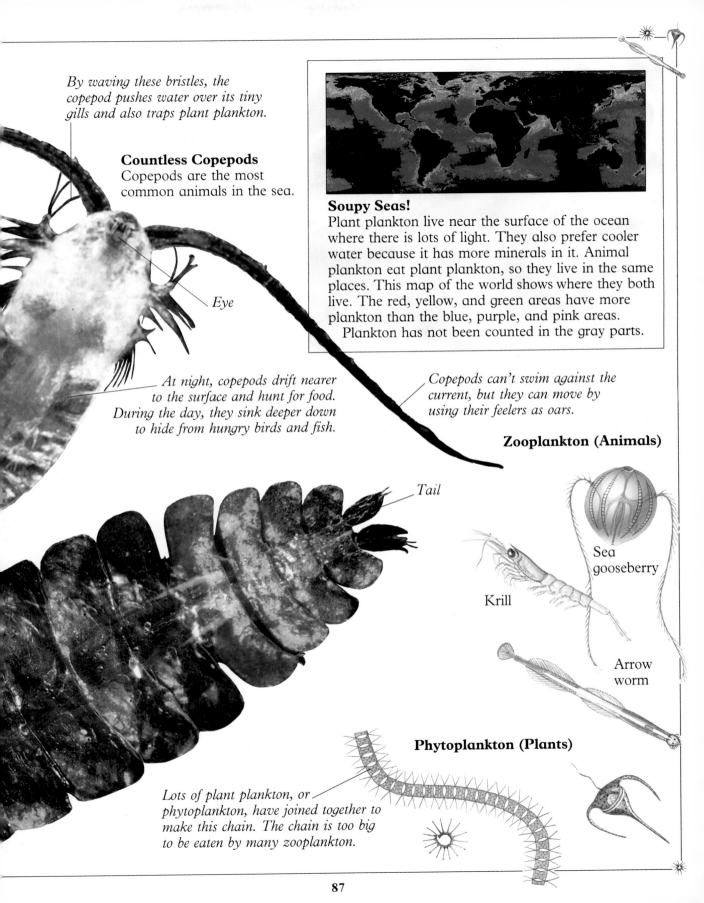

Eye

At night, copepods drift nearer to the surface and hunt for food. During the day, they sink deeper down to hide from hungry birds and fish.

Soupy Seas!
Plant plankton live near the surface of the ocean where there is lots of light. They also prefer cooler water because it has more minerals in it. Animal plankton eat plant plankton, so they live in the same places. This map of the world shows where they both live. The red, yellow, and green areas have more plankton than the blue, purple, and pink areas. Plankton has not been counted in the gray parts.

Copepods can't swim against the current, but they can move by using their feelers as oars.

Zooplankton (Animals)

Tail

Sea gooseberry

Krill

Arrow worm

Phytoplankton (Plants)

Lots of plant plankton, or phytoplankton, have joined together to make this chain. The chain is too big to be eaten by many zooplankton.

SHELLFISH

The shells that you find on a beach are the empty homes of small animals called shellfish. These soft-bodied creatures need hard shells to protect them from starfish, crabs, fish, and even birds. The most common types of shellfish are gastropods and bivalves. Gastropods are underwater snails. They grow coiled shells and slide around on a slimy foot. Bivalves have two, flatter shells that cover their whole bodies.

Growing Up
Unlike crabs, shellfish never need new shells. Baby shellfish, or larvae, grow a tiny shell that keeps on getting bigger.

Eggs

Larva

Bubble shells are able to use their huge mantles to swim, because they have very thin, light shells.

Mantle

The Animal Inside
Bivalves, gastropods, and all other shellfish have soft bodies and no backbones. They are mollusks, just like octopuses!

Worm

These feelers scan the water for food.

Water in

Water out

Swimming Scallops
By taking in water and then shooting it out its back end, scallops propel themselves through the sea.

Eye

Buried Alive
Many bivalves spend most of their lives buried in sand. They dig a hole with their foot, then poke tubes, or siphons, into the sea. The siphons take water into the gills and catch food.

Siphon

Foot

Tusk shell

Tellin

Sand gaper

Razor shell

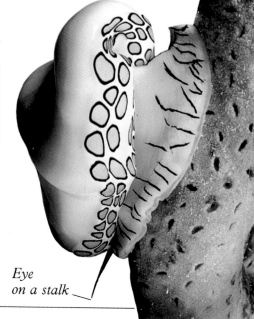

Eye on a stalk

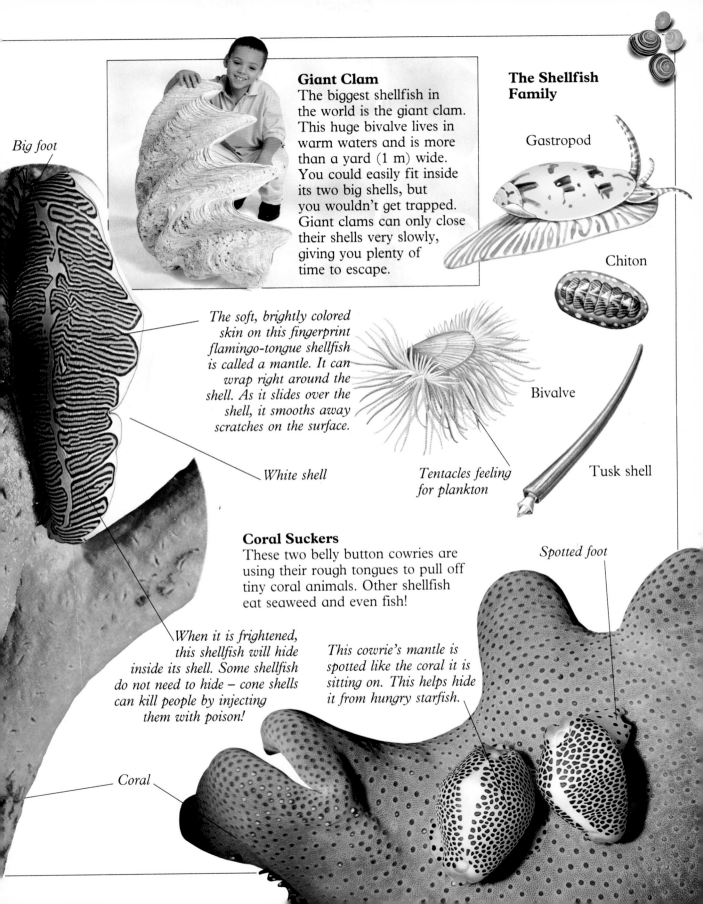

Big foot

Giant Clam
The biggest shellfish in the world is the giant clam. This huge bivalve lives in warm waters and is more than a yard (1 m) wide. You could easily fit inside its two big shells, but you wouldn't get trapped. Giant clams can only close their shells very slowly, giving you plenty of time to escape.

The Shellfish Family

Gastropod

Chiton

The soft, brightly colored skin on this fingerprint flamingo-tongue shellfish is called a mantle. It can wrap right around the shell. As it slides over the shell, it smooths away scratches on the surface.

Bivalve

White shell

Tentacles feeling for plankton

Tusk shell

Coral Suckers
These two belly button cowries are using their rough tongues to pull off tiny coral animals. Other shellfish eat seaweed and even fish!

Spotted foot

When it is frightened, this shellfish will hide inside its shell. Some shellfish do not need to hide – cone shells can kill people by injecting them with poison!

This cowrie's mantle is spotted like the coral it is sitting on. This helps hide it from hungry starfish.

Coral

CRABS

What lives in the sea or on land, can be any color, has its eyes on stalks, swims and walks sideways, and carries its own house? Would you have guessed a crab? Crabs live in all parts of the sea, from the very deepest oceans to wave-swept shores.

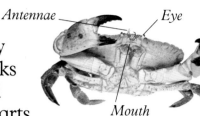

Antennae — — Eye

Mouth

Crabs use their rear four pairs of legs to scuttle sideways.

Its strong claws are used for fighting and also for tearing apart fish, shellfish, and plants to eat.

Under this spine, there are two feelers, called antennae. The crab uses tiny hairs along the antennae to touch, smell, and taste.

Its shell is often called a carapace.

When crabs get too big for their shells, they split them open and shed them. Underneath the old suit of armor, there is a new, soft shell. This can take three days to harden.

Crabs can grow new legs if they are torn off.

This spiny spider crab is protected from most of its enemies by its shell.

Crab Eaters
Crabs make a tasty meal for fish, birds, octopuses, seals, and people!

Crabs breathe through five pairs of gills. These are inside the shell, near the top of each leg.

Joint

90

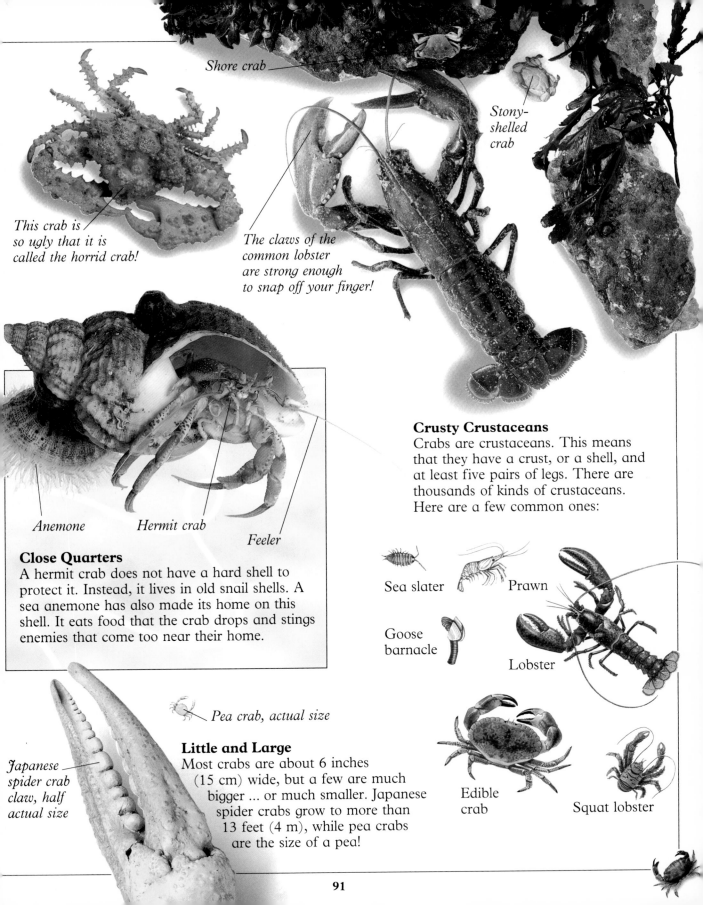

Shore crab

Stony-shelled crab

This crab is so ugly that it is called the horrid crab!

The claws of the common lobster are strong enough to snap off your finger!

Anemone

Hermit crab

Feeler

Close Quarters

A hermit crab does not have a hard shell to protect it. Instead, it lives in old snail shells. A sea anemone has also made its home on this shell. It eats food that the crab drops and stings enemies that come too near their home.

Crusty Crustaceans

Crabs are crustaceans. This means that they have a crust, or a shell, and at least five pairs of legs. There are thousands of kinds of crustaceans. Here are a few common ones:

Sea slater

Prawn

Goose barnacle

Lobster

Pea crab, actual size

Little and Large

Most crabs are about 6 inches (15 cm) wide, but a few are much bigger ... or much smaller. Japanese spider crabs grow to more than 13 feet (4 m), while pea crabs are the size of a pea!

Japanese spider crab claw, half actual size

Edible crab

Squat lobster

STARFISH

Starfish are star-shaped, but they are not fish – they are echinoderms. This means that they have spiny skin. They cannot swim, but they are very good at crawling! They can walk up seaweed fronds and climb down the sides of rocks. Even in the deepest, darkest parts of the sea, there are starfish creeping around.

This crimson-knobbed starfish, like most starfish, measures less than eight inches (20 cm) from tip to tip. But some species are as wide as a small car!

Starfish don't have eyes. Instead, they have eyespots on the tips of their arms. These special cells cannot see shapes, but they can tell whether it is light or dark.

These bumps are actually spines.

Burrowing starfish

A starfish's arms are very flexible. This is because its skeleton is made up of lots of tiny spines that can move in any direction.

Central disc

Cushion star

Starfish breathe through their feet and also through tiny tubes that are found all over their bodies.

Common starfish

Stomach This
Starfish eat shellfish. When a starfish finds a clam, it pulls open its shell with its tiny tube feet, pushes its whole stomach inside the shell, and then slowly digests the clam.

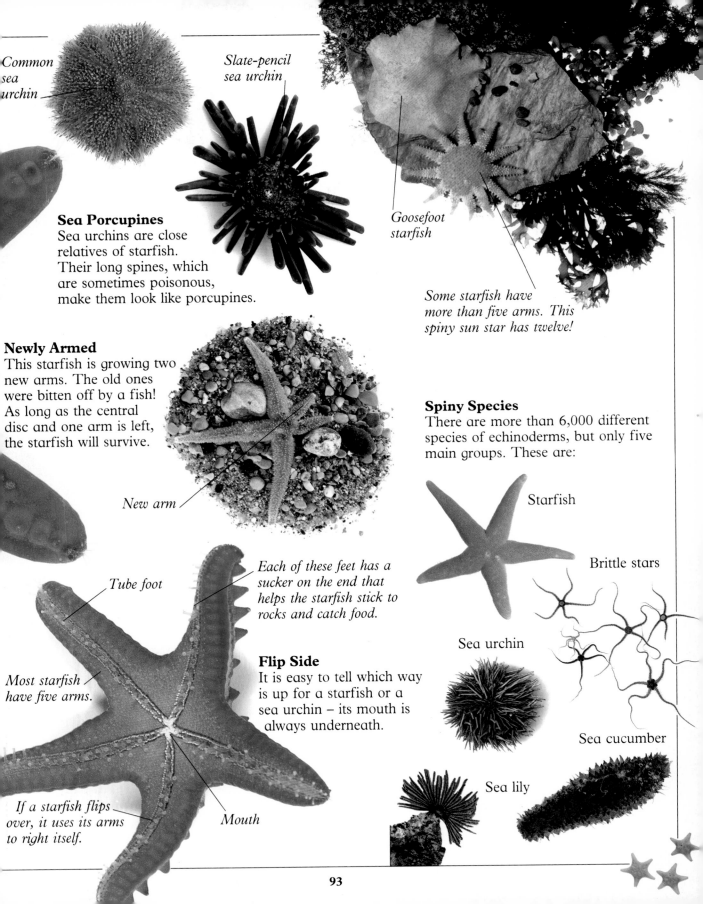

*Common
sea
urchin*

*Slate-pencil
sea urchin*

*Goosefoot
starfish*

Sea Porcupines
Sea urchins are close
relatives of starfish.
Their long spines, which
are sometimes poisonous,
make them look like porcupines.

*Some starfish have
more than five arms. This
spiny sun star has twelve!*

Newly Armed
This starfish is growing two
new arms. The old ones
were bitten off by a fish!
As long as the central
disc and one arm is left,
the starfish will survive.

New arm

Spiny Species
There are more than 6,000 different
species of echinoderms, but only five
main groups. These are:

Starfish

Brittle stars

Tube foot

*Each of these feet has a
sucker on the end that
helps the starfish stick to
rocks and catch food.*

Sea urchin

Flip Side
It is easy to tell which way
is up for a starfish or a
sea urchin – its mouth is
always underneath.

*Most starfish
have five arms.*

Sea cucumber

Sea lily

*If a starfish flips
over, it uses its arms
to right itself.*

Mouth

OCTOPUSES

Did you know that octopuses are related to snails? But unlike most other mollusks, they don't have shells to protect them. Instead, these eight-armed animals squeeze their soft bodies into small cracks or holes in rocks. Once they are safely hidden, it is very hard for conger eels, sharks, seals, and people to find and eat them.

Bunch of Eggs
Inside each of these soft-shelled eggs, there is a baby octopus.

Octopuses can see shapes and colors very well with their large eyes.

Gone Fishing
Octopuses hunt for their food. They pounce on fish, starfish, and crabs. Some have webs between their arms that help them net even more animals.

Tentacle

This common octopus is about four inches (10 cm) wide. The largest octopus ever found was more than 29 feet (9 m) from tip to tip!

Web

An octopus can shoot ink, called sepia, out of its siphon. This black cloud hangs in the water and hides the octopus from its enemies.

Siphon

Jet-propelled
If an octopus is frightened, it does not crawl slowly away – it jets off! By forcing water out through its siphon, it can shoot through the water.

Suckers
Rows of super-strong suckers help octopuses hang on to rocks, grab food, and pull themselves along the ocean floor.

The Cephalopod Family
Octopuses and their relatives are known as cephalopods. This means they are mollusks that live in the ocean and have tentacles.

Cuttlefish

Vampire squid

In less than a second, an octopus can change color.

The soft body of an octopus is like a big bag of skin. Water flows into this stretchy bag, passes over the gills, and then escapes through a special funnel, called a siphon.

Squid

Sucker

Nautilus

Battle of the Giants
There are stories of huge octopuses swallowing people whole, but this doesn't actually happen. These myths may be based on giant squid, which can be 65 feet (20 m) long and are close relatives of octopuses. Deep down in the sea, giant squid are believed to battle against sperm whales.

Sperm whale

Giant squid

Octopus

FISH

All living things need a gas called oxygen to breathe. You cannot see it, but it is found in air and water. You use your lungs to breathe in air. If you swim under water, you either have to hold your breath or use a snorkel. Fish don't have to do this. They can take their oxygen straight out of the water.

Fish Eggs
Most fish lay jellylike eggs. Some guard their eggs until they hatch. Others, like cod, just squirt millions of tiny eggs into the sea. This is called spawning.

Dorsal fin

Fish do not need eyelids – seawater keeps their eyes wet and free of dirt.

Water in

Water out   *Gills*

Breathing Under Water
Water flows into a fish's mouth, over flaps called gills, and out through the gill openings. The gills take the oxygen out of the water.

Pectoral fin

There are four gills behind this opening.

Bony fish have a "balloon" inside them called a swim bladder. The air in this organ helps fish stay afloat.

Its pectoral and pelvic fins help a fish move up, down, left, or right.

Pelvic fin

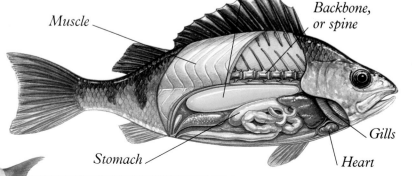

Muscle

Backbone, or spine

Gills

Stomach

Heart

Inside Story
Most of the important parts of a bony fish are in the lower half of its body. The top half is full of muscles that move its tail.

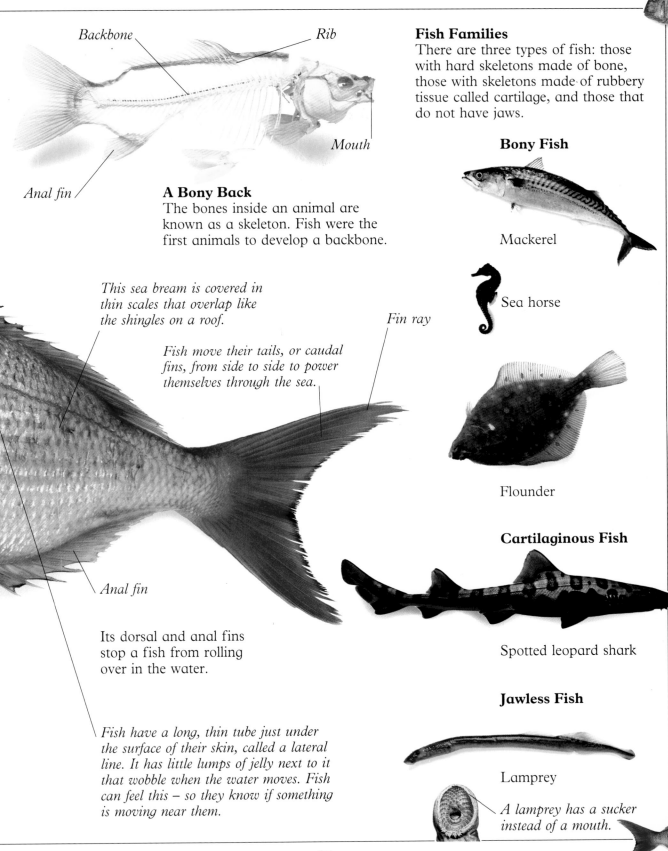

Backbone

Rib

Mouth

Anal fin

A Bony Back
The bones inside an animal are known as a skeleton. Fish were the first animals to develop a backbone.

This sea bream is covered in thin scales that overlap like the shingles on a roof.

Fish move their tails, or caudal fins, from side to side to power themselves through the sea.

Fin ray

Anal fin

Its dorsal and anal fins stop a fish from rolling over in the water.

Fish have a long, thin tube just under the surface of their skin, called a lateral line. It has little lumps of jelly next to it that wobble when the water moves. Fish can feel this – so they know if something is moving near them.

Fish Families
There are three types of fish: those with hard skeletons made of bone, those with skeletons made of rubbery tissue called cartilage, and those that do not have jaws.

Bony Fish

Mackerel

Sea horse

Flounder

Cartilaginous Fish

Spotted leopard shark

Jawless Fish

Lamprey

A lamprey has a sucker instead of a mouth.

SHARKS

Sharks are some of the best hunters in the ocean. With their razor-sharp teeth and huge jaws, they can tear up seals, fish and even wooden boats! Tiger sharks and great white sharks sometimes attack people, but most sharks are frightened by people and avoid them.

See how big this jaw is!

Sharks are like a swimming nose. They can smell injured animals and other food that is hundreds of yards away.

All living animals produce a small amount of electricity. You cannot sense it, but sharks can. Little dimples on their heads work like television antennae to tell them where animals are hiding.

Five gill slits on each side of the head let water into the shark's gills.

Teeth as Sharp as Knives
A shark's teeth can cut through skin and crunch up bones, but they soon get blunt. Each tooth only lasts for a few weeks, then it falls out and is replaced by a new one. Basking sharks eat plankton, so they don't have any teeth!

Dorsal fin

Pectoral fin

Super Swimmer
A shark swims by bending from side to side. First it moves its head, then its body, and last of all its long tail. As this wave travels down its body, it pushes the shark through the sea.

Shark Attack!
Just before they bite, sharks bend their noses up and thrust their teeth forward.

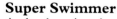

Great white shark

Sandtiger shark

Blue shark

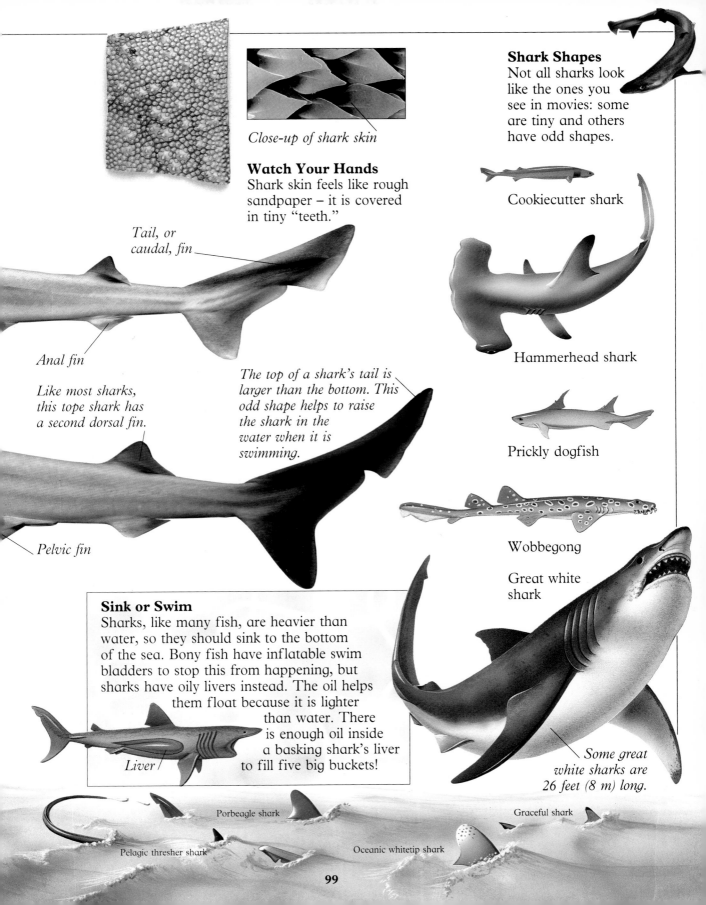

Close-up of shark skin

Shark Shapes
Not all sharks look like the ones you see in movies: some are tiny and others have odd shapes.

Cookiecutter shark

Watch Your Hands
Shark skin feels like rough sandpaper – it is covered in tiny "teeth."

Hammerhead shark

Tail, or caudal, fin

Anal fin

Like most sharks, this tope shark has a second dorsal fin.

The top of a shark's tail is larger than the bottom. This odd shape helps to raise the shark in the water when it is swimming.

Prickly dogfish

Wobbegong

Pelvic fin

Great white shark

Sink or Swim
Sharks, like many fish, are heavier than water, so they should sink to the bottom of the sea. Bony fish have inflatable swim bladders to stop this from happening, but sharks have oily livers instead. The oil helps them float because it is lighter than water. There is enough oil inside a basking shark's liver to fill five big buckets!

Liver

Some great white sharks are 26 feet (8 m) long.

Pelagic thresher shark

Porbeagle shark

Oceanic whitetip shark

Graceful shark

RAYS

Rays are some of the most common fish in the ocean. A few look like their relatives, the sharks, but most of them are diamond-shaped and flat. Some of these shy, gentle animals can be glimpsed gliding through the sea, but others spend their time lying on the ocean floor.

These sharp spines protect the ray. Stingrays also have one very large poisonous spine that they use to stab their enemies.

Like sharks, rays have skeletons made of cartilage.

Gill slit

Mouth

Thornback ray

Upside Down
On the underside of a ray, there are ten gill slits and a mouth that is full of flat teeth.

Small pelvic fin

Mighty Manta
Most rays are less than two yards (1.9 m) wide but a manta ray can grow to more than eight yards (7.3 m) in width. Luckily for divers, it only eats plankton and small fish.

A ray's gill slits are on the underside of its body, so when it is lying on the sea bottom, its gills are blocked. When this happens, clean water enters through a tiny hole behind its eyes, called a spiracle.

By flapping their fins, rays disturb the sand and uncover crabs – a tasty snack for a ray.

Underwater Flying
A ray is one of the most graceful swimmers in the sea. Flapping its huge fins, it "flies" like a giant underwater bird.

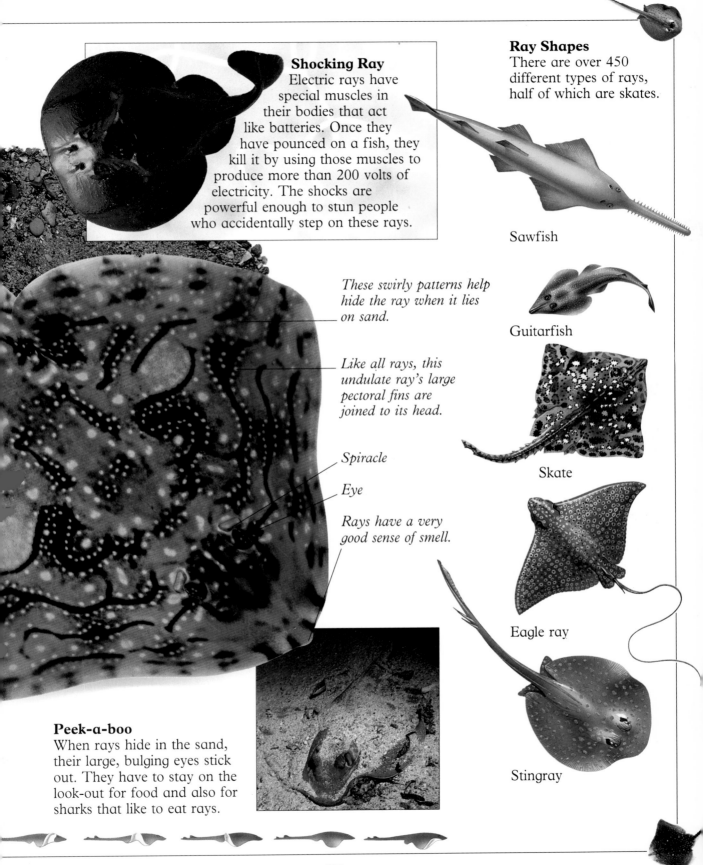

Shocking Ray

Electric rays have special muscles in their bodies that act like batteries. Once they have pounced on a fish, they kill it by using those muscles to produce more than 200 volts of electricity. The shocks are powerful enough to stun people who accidentally step on these rays.

Ray Shapes

There are over 450 different types of rays, half of which are skates.

Sawfish

Guitarfish

Skate

Eagle ray

Stingray

These swirly patterns help hide the ray when it lies on sand.

Like all rays, this undulate ray's large pectoral fins are joined to its head.

Spiracle

Eye

Rays have a very good sense of smell.

Peek-a-boo

When rays hide in the sand, their large, bulging eyes stick out. They have to stay on the look-out for food and also for sharks that like to eat rays.

WHALES

Millions of years ago whales used to walk. Since then, they have changed a lot. They have grown bigger, their back legs have disappeared, and their front legs have turned into flippers. They can't live on land anymore, but they are still mammals. This means they breathe air and feed their babies milk.

There are two kinds of whales: toothed whales and toothless whales, called baleen whales.

Krill for Lunch
Baleen whales eat krill, a kind of plankton. A large whale can eat two tons a day – half the weight of an elephant!

The whale moves its tail up and down to push itself through the sea. The fastest whale is the sei. It can reach speeds of 30 miles an hour (48 km/h), ten times faster than a person can swim.

No animal has longer flippers than a humpback whale – they are about five yards (5 m) long, almost as tall as a giraffe!

There She Blows!
When a whale surfaces to fill its lungs with fresh air, warm air escapes from its blowhole. This escaping air mists up, just like your breath on a cold day, and forms a tall spout called a blow.

Deep throat grooves let a baleen whale's mouth stretch to hold vast amounts of seawater and krill.

Whales have thick fat, called blubber.

Jumping for Joy
Humpback whales love to leap right out of the water. This is called breaching. When they crash back into the ocean, they make a huge splash!

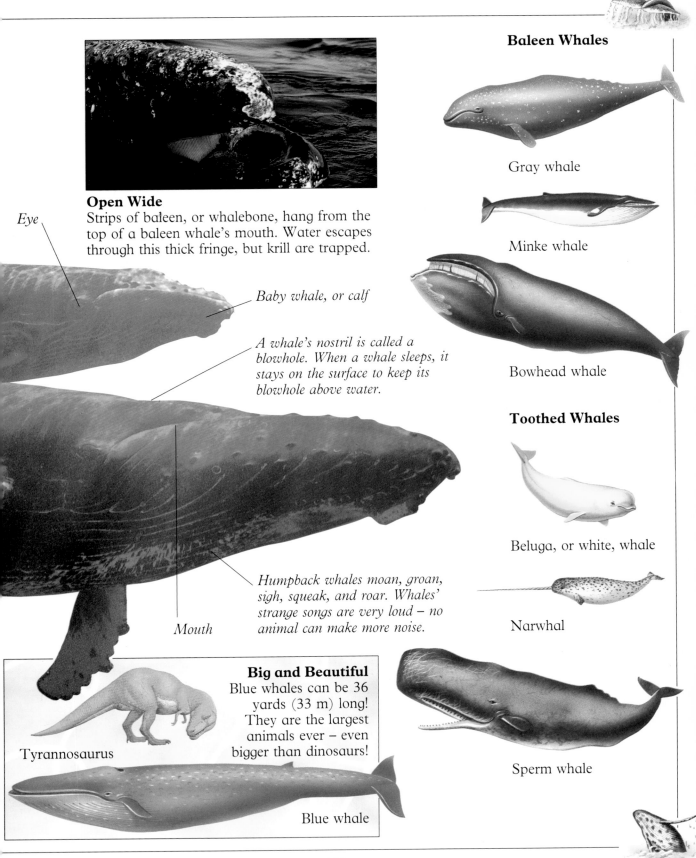

Baleen Whales

Gray whale

Minke whale

Bowhead whale

Toothed Whales

Beluga, or white, whale

Narwhal

Sperm whale

Open Wide
Strips of baleen, or whalebone, hang from the top of a baleen whale's mouth. Water escapes through this thick fringe, but krill are trapped.

Eye

Baby whale, or calf

A whale's nostril is called a blowhole. When a whale sleeps, it stays on the surface to keep its blowhole above water.

Humpback whales moan, groan, sigh, squeak, and roar. Whales' strange songs are very loud – no animal can make more noise.

Mouth

Big and Beautiful
Blue whales can be 36 yards (33 m) long! They are the largest animals ever – even bigger than dinosaurs!

Tyrannosaurus

Blue whale

DOLPHINS

Dolphins are small, toothed whales. Some people think that these smooth-skinned mammals are highly intelligent. They learn quickly and seem to talk to one another with whistles, clicks, and grunts. Since ancient times, there has been a special friendship between humans and these playful animals. There are many stories of dolphins saving drowning sailors.

Large, curved dorsal fin

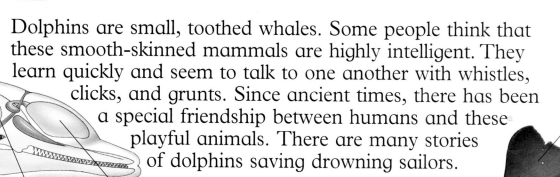

Brain

Melon

Built-in Sonar
In the dark, dolphins use sound to find their food and to figure out where they are. The melon, a fatty area in the head, sends out clicks that bounce back when they hit something. These echoes tell dolphins where an object is.

Dolphins do not need to drink. They get all the water they need from the fish and squid they eat.

Sonar clicks

The shape of a dolphin's mouth makes it look as if it is always smiling.

Toothy Grin
Bottlenose dolphins have more than 100 teeth, each about one-third of an inch (8 mm) long.

Hitching a Ride
Just in front of a speeding ship, there is a wave. Dolphins love to swim in this spot. Like surfers, if they catch the wave in the right way, it will push them through the sea.

Blowhole

Most dolphins have a long snout, called a beak.

Family Life

Dolphins live in family groups called herds. Baby dolphins often stay with their mothers for many years. They learn how to catch fish, signal to each other, and escape from sharks by copying other members of their herd.

Harbor porpoise

Bottlenose dolphin

Dolphins, like all toothed whales, have only one blowhole. Baleen whales have two nostrils.

Bottlenose dolphins are very playful. They love to leap out of the water.

Striped dolphin

Common dolphin

Tail fluke

Just as you have your own personal name and voice, every dolphin has its own whistle, which other dolphins recognize.

Risso's dolphin

Killer whale (male)

The Biggest Dolphin

Killer whales are fierce dolphins that can grow to be 30 feet (9 m) long. This one is trying to grab a sea lion.

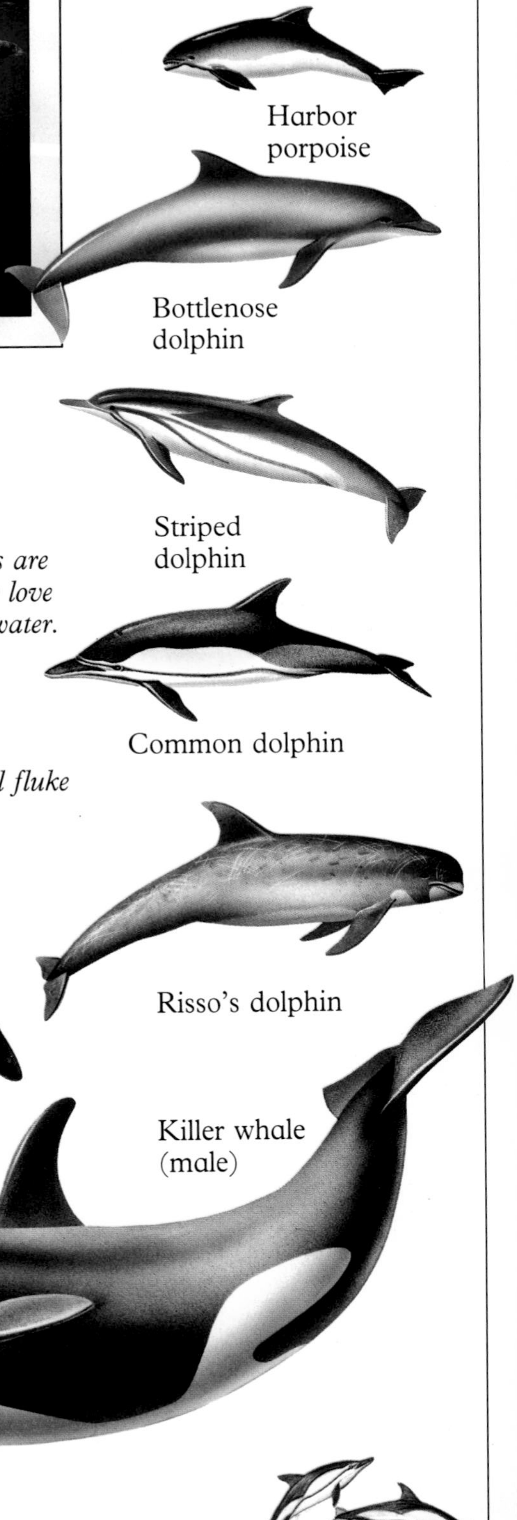

SEALS

Seals are warm-blooded mammals, which means that they can make their own heat. Because they often swim in very cold water, they need to be able to keep this heat inside their bodies. Seals can't put on a warm coat like you do to keep warm. Instead, they are covered with short hair or fur and have a layer of fat, called blubber. Sometimes, when seals swim in warmer water, they get too hot and have to fan their flippers in the air to cool down.

Ball of Fluff
Baby seals, or pups, are born on land. For a few weeks, many of them have white, fluffy fur.

All seals have good hearing, but only sea lions and fur seals have ear flaps. All that can be seen of this harbor seal's ears are two tiny holes.

When a seal dives under water, it closes its nose, mouth, and ears!

These whiskers, which are 40 times thicker than human hair, can sense movement in the water. This helps the seal find fish, shellfish, squid, and octopus to eat.

Pile Up!
When walruses climb up onto beaches, they often lie right on top of each other to keep warm!

Flip-flop
Sea lions, fur seals, and walruses use their front flippers to sit up straight, and their back flippers can turn forward. This means that they can walk, and even run, on dry land. Hair seals can only slide around on their bellies when they leave the sea.

The Seal Family
There are three different groups of seals:

Gray seal

Hair seals

Elephant seal

Its smooth, streamlined body helps the seal speed away from killer whales and polar bears.

Eared seals

Northern fur seal

California sea lion

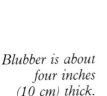

Blubber is about four inches (10 cm) thick.

Seal-eating Seals
Leopard seals are fierce. They leap out of the sea and thump onto the ice to grab penguins or other seals.

This seal is less than two yards (2 m) long. An elephant seal can measure nearly six yards (6 m) from nose to back flippers, which is more than the height of a giraffe!

Walruses

Male walruses fight each other with their big teeth.

Hair seals, like this harbor or common seal, speed through the ocean by moving their back flippers up and down.

INSECTS AND SPIDERS

Millions of years ago, long before dinosaurs arrived, there were insects on Earth. Scientists have discovered more than a million types of insects, and there are many more waiting to be found.

All insects start as eggs, and most go through a larval stage. Their bodies are divided into three parts – head, thorax, and abdomen. Adults have six legs, and most have wings.

Spiders are not insects – they are arachnids. In fact, spiders love to eat insects. Unlike an insect, a spider has eight legs, and its head and thorax are joined.

Both insects and spiders live hidden away, creeping through undergrowth and living in gardens and forests – but occasionally, they come to stay in our homes.

Garden spider

Dragonfly

Honeybee

Peacock butterfly

Peacock butterfly caterpillar

Cactus longhorn beetles

Red ants

Grasshopper

Common
housefly

109

FLIES

You have probably seen a housefly zooming around your kitchen. Most people think of flies as pests – annoying little creatures that buzz around us, bite us, walk on our food, and spread disease. But in other ways, flies are a necessary and useful part of our world. They help to pollinate plants and are eaten by a variety of other animals.

A hover fly taking off

Sensitive Flies
Flies have surprisingly strong senses. This means they have very good eyesight and keen senses of taste and smell.

Flies have two large compound eyes, which see colors and shapes.

Many flies, like the housefly, have small antennae.

When the fly finds liquid food, it simply sucks it up. If it finds solid food, the fly first dissolves it with special juices.

Nice to See You!
A human eye has just one rounded lens. A housefly has thousands of six-sided lenses, and each lens sees a part, or piece, of a bigger picture. When a housefly looks at you, it sees you as different pieces.

At the end of a fly's mouth are two pads that look like lips.

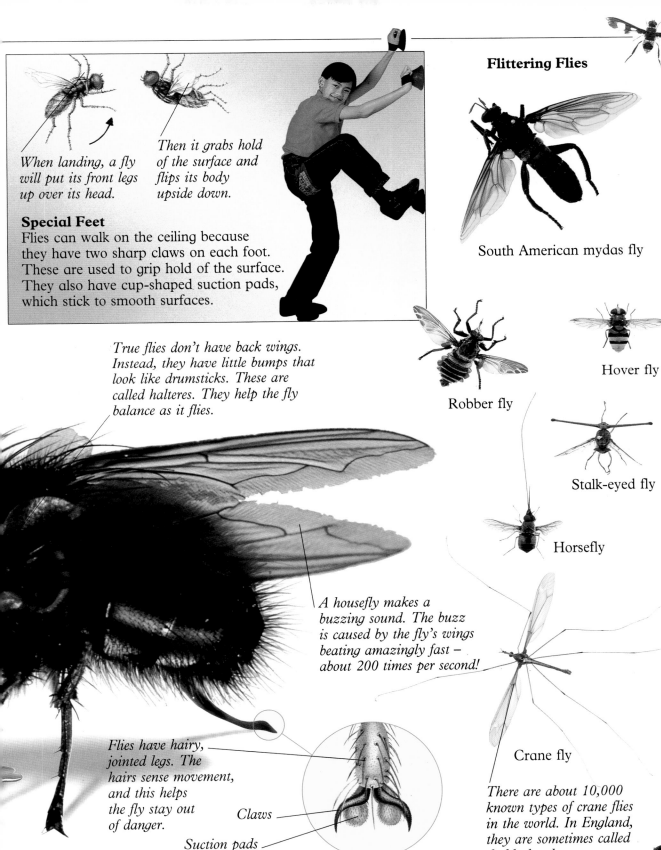

When landing, a fly will put its front legs up over its head.

Then it grabs hold of the surface and flips its body upside down.

Special Feet
Flies can walk on the ceiling because they have two sharp claws on each foot. These are used to grip hold of the surface. They also have cup-shaped suction pads, which stick to smooth surfaces.

True flies don't have back wings. Instead, they have little bumps that look like drumsticks. These are called halteres. They help the fly balance as it flies.

A housefly makes a buzzing sound. The buzz is caused by the fly's wings beating amazingly fast – about 200 times per second!

Flies have hairy, jointed legs. The hairs sense movement, and this helps the fly stay out of danger.

Claws

Suction pads

Flittering Flies

South American mydas fly

Robber fly

Hover fly

Stalk-eyed fly

Horsefly

Crane fly

There are about 10,000 known types of crane flies in the world. In England, they are sometimes called daddy longlegs.

BEETLES

There are more species of beetles in the world than any other kind of animal. It is thought there are at least 300,000. Most are plant-eaters, but some battling beetles attack and eat other insects and are quite ferocious. Beetles can be pests because they eat valuable crops. But mostly they are helpful to us because they eat dead plants and animals and return them to the soil as important nutrients.

Ready for Battle
A beetle's hard outer casing acts like protective armor.

Forewing

Take That!
These fighting beetles are stag beetles. They get their name because male stag beetles have large "horns." These are really large jaws that are used for fighting like the antlers of a real stag. Males fight to defend their territory.

The tough, black wing case protects the delicate hind wings.

Hind wing

Special Friend
Ladybugs are beetles, too. Not all beetles can fly, but ladybugs can. They use their hind wings to fly.

The beetle's claws help it grip.

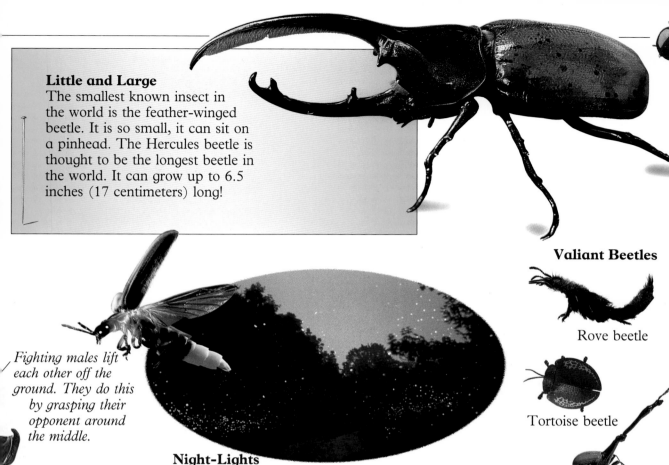

Little and Large

The smallest known insect in the world is the feather-winged beetle. It is so small, it can sit on a pinhead. The Hercules beetle is thought to be the longest beetle in the world. It can grow up to 6.5 inches (17 centimeters) long!

Valiant Beetles

Rove beetle

Tortoise beetle

Giraffe beetle

Bombardier beetle

Fighting males lift each other off the ground. They do this by grasping their opponent around the middle.

Night-Lights

Fireflies are not really flies. They are beetles. At night, the females put on a light show as they flash their tails to attract a mate. They are able to do this because they have a special chemical in their bodies.

Hard, antlerlike jaws

If it's attacked, it fires off a mixture of burning chemicals.

Beetles have palps to help them sense food.

Roll Over

Dung beetles go to a lot of trouble to find a safe and nutritious home for their young. They collect animal dung and roll it into large balls. They roll the dung balls all the way to their underground homes. There they lay their eggs in the dung. When the beetle larvae hatch, they discover a tasty meal in front of them!

Beetle storing dung

CATERPILLARS

Caterpillars are like tiny eating machines. They spend most of their time chomping on leaves. Caterpillars are actually the young of butterflies and moths. They hatch from the eggs the adult female has laid on plants. With constant eating, they get bigger and bigger, until they are ready to change into butterflies and moths.

A New Skin
Caterpillar skin cannot stretch. So as it gets larger, the caterpillar breaks out of its skin. Underneath is a new, larger skin, which will last until the caterpillar needs to molt again.

The front three pairs of legs are called thoracic legs and are used for walking and clasping.

Caterpillars produce silk from special glands and force it out through a spinneret under the head.

A caterpillar has 12 tiny, simple eyes, called ocelli, on its head.

Hatching Out
The butterfly's eggs are laid on the underside of a leaf.

The egg gets darker as the caterpillar prepares to hatch.

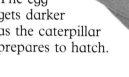

The caterpillar bites its way out and then pulls itself free of the eggshell.

The caterpillar's head is armed with a pair of stout jaws, called mandibles.

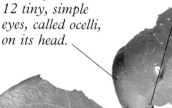

Caterpillars test food and guide it to their mouth with mouthparts, called maxillary palps.

The body is divided into 13 segments.

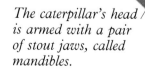

The first meal is the eggshell, which is full of nourishment itself.

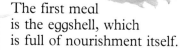

Wonderful Caterpillars

Caterpillars can be hairy or spiny and have unusual shapes.

Tiger moth caterpillar

Cabbage white caterpillar

Emperor moth caterpillar

Puss moth caterpillar

Like most insects, caterpillars breathe through openings called spiracles.

The five pairs of stumpy, suckerlike legs are called prolegs. The caterpillar uses them for clinging onto plant stalks.

Silkworms

Silk is produced by most moth caterpillars. But the finest silk is produced by the large white moth caterpillar, often known as a silkworm. After the caterpillars have spun themselves into a silken cocoon, they are put into boiling water. The silk is removed and spun into threads to create material for clothes.

Masters of Disguise

To avoid being eaten, some caterpillars have developed crafty disguises. The hawk-moth caterpillar looks like a deadly snake, the lobster moth looks like a raised lobster's claw, and the common sailer looks like a shriveled leaf.

Hawkmoth

Lobster moth

Common sailer

Excuse Me, I'm Changing

To become adult moths, most moth caterpillars spin themselves a cocoon using silk, which comes out of the spinneret. Inside, they undergo astonishing physical changes.

BUTTERFLIES

Butterflies are perhaps the most beautiful of all insects. It is amazing to think that a fat, leaf-eating caterpillar can become a brightly colored, fluttering creature of the air. The change happens in the butterfly chrysalis. The caterpillar's body is broken down and completely changed. After about four weeks, a fully formed butterfly emerges.

Time to Wake Up
The butterfly comes out of the chrysalis in three stages. During this time, it is very open to attack by hungry birds or spiders.

1. No longer a caterpillar, a beautiful butterfly comes out of the chrysalis with its wings crumpled up.

Monarchs on the Move
Most butterflies hatch, live, and die in one place. But when winter comes to the eastern and western coasts of North America, thousands of monarch butterflies head south to the warmth of California and Mexico. When warm weather returns to their first home, they fly north again.

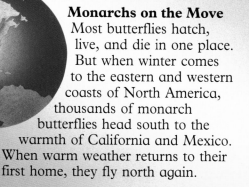

Happy Landings

[A] clouded yellow butterfly comes [t]o land on a thistle. Butterfly [fli]ght is more controlled than it [loo]ks. The insect is able to [ch]ange course instantly and [m]ake sudden landings.

Butterflies feed through a tube, called a proboscis. This is coiled up when not in use.

Butterflies have clubbed antennae.

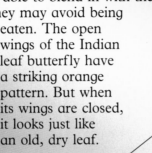

Wings open

In the Background

Butterflies make a tasty meal for birds. But if they are able to blend in with their background, they may avoid being eaten. The open wings of the Indian leaf butterfly have a striking orange pattern. But when its wings are closed, it looks just like an old, dry leaf.

Wings closed, resting on leaf

Brilliant Butterflies

Male bird-wing butterfly

Glass swallowtail butterfly

88 butterfly

Cramer's blue morpho butterfly

2. The butterfly must stay still for some hours as blood is pumped into the wing veins to stretch the wings. Later, it holds its wings apart to let them harden.

3. When its wings have hardened, the butterfly is ready to fly off to find its first meal of nectar.

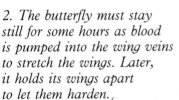

Scaly Wings

The wings of both butterflies and moths are made up of tiny scales, which overlap like the tiles on a roof. Bright colors can be used either to attract a mate or to warn predators that the butterfly or moth is not good to eat.

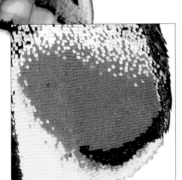

MOTHS

Most moths are night fliers, and their strong senses of smell and hearing make them well suited to a nighttime existence. They can easily find their way through darkness, and although attracted to light, they are dazed by it. Moths rest by day, and many are colored to look like tree bark or leaves so that they cannot be spotted by natural enemies, such as birds and lizards.

Moths' antennae are straight or fernlike. They are used for smelling out nectar, or other moths at night.

The South American ghost moth has the biggest wingspan of any moth. Wing tip to wing tip, it can measure up to 12 inches (30 cm).

Nymphalid butterfly

Wings at Rest
One way to tell a moth from a butterfly is to see how the insect folds its wings. Butterflies close their wings upright. But most moths rest with their wings folded over their backs.

White ermine moth

A moth's body is thick and strong.

A moth's wings are joined together.

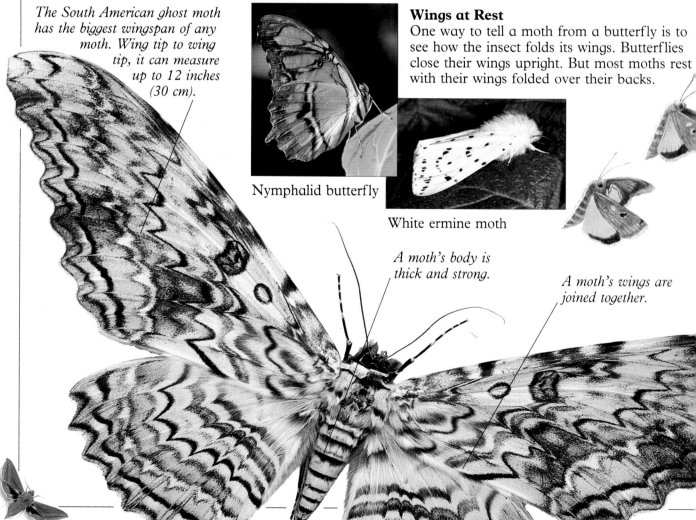

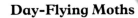

Whooo's Whooo?

To scare off enemies, such as birds, the wing patterns of some moths mimic the appearance of fierce animals. The great peacock moth has big eyespots on its wings, which look like an owl's eyes. With these staring back at them, birds think twice before attacking!

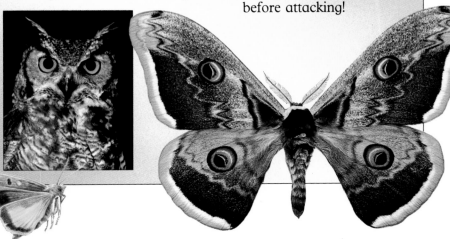

Colombian blue-wing moth

Madagascan red-tailed moth

Verdant hawkmoth

Sloan's uraniid moth

A Real Eyeful!

Pyraustine moths have strange feeding habits. With their long proboscises, they drink the tears of animals such as cows and buffaloes. They are so gentle that the animal's eye does not become irritated.

The veins of the moth's wing help warm or cool the insect.

Spot the Moth!

This geometrid moth from the jungles of Borneo looks like lichen on a tree trunk. The secret of its camouflage is not just color, but also ragged outlines and broken patterns.

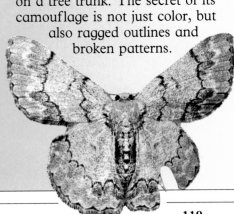

ANTS

Ants live together in nests that are like underground towns. There may be up to half a million ants in one nest. Most of these are female and are called workers. Some workers build and repair the nest, while some are "soldiers" and guard the entrance. Others gather food for the larvae and the huge queen. Her life is spent laying millions of eggs, and the survival of the nest depends upon her well-being.

Weaving Away
Weaver ants make their nests in trees. They sew leaves together using a sticky silk thread produced by their larvae. The queen lives inside the leaf envelope.

Some ants will spray a nasty chemical from their rear end if they sense danger!

Ants have powerful jaws, or mandibles, for chopping food. The mouth is just below the mandibles.

When ants meet, they "tap" antennae. The antennae contain chemical "messages" that can be passed on by touching.

Ants can run very fast because they have long legs.

Farming Fungus
Leaf-cutting ants are the farmers of the ant world. They cut up bits of leaves and take them back to the nest. Fungus grows on the rotting leaves – and then the ants feed on the fungus!

Left, Right!

Driver ants from South America are very fierce insects. They are nomadic, which means they are always on the move. They march in columns through the forest, killing and eating everything in their path. Here they are raiding a wasp's nest.

Bustling Ants

Red ant

Black ant

Harvester ant

Wood ant

Dinoponera –
the largest ant

Living Larders

Some honey-ant workers spend their whole lives feeding on nectar. Their abdomens swell. Then, when food is hard to find, other workers use them as a food supply.

Inside an Ants' Nest

The success of an ants' nest relies on the hardworking and organized inhabitants.

The queen lays her eggs in the royal chamber.

Soldier ant guarding the entrance

Seed-crushing ant

Wingless workers take care of the ant larvae.

A network of tunnels joins the chambers in the ants' nest.

GRASSHOPPERS

Grasshoppers are known for the "ticking" sounds they make and for their ability to leap high into the air. There are more than 20,000 different kinds of grasshoppers in the world. Grasshoppers are plant-eaters, feeding on leaves and stems. Normally, they prefer to be alone. But under special conditions, they undergo a series of physical changes. They increase in size, become more brightly colored, and gather in millions to become a swarm of hungry locusts.

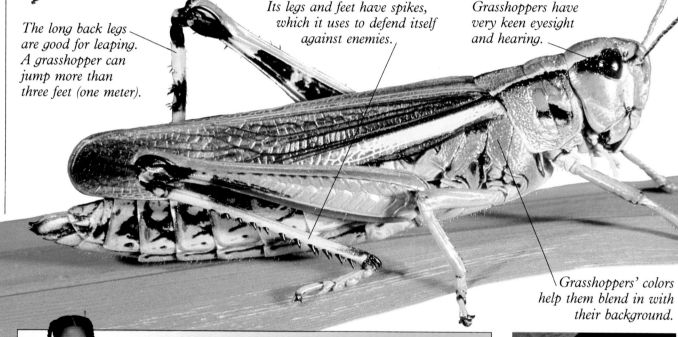

The long back legs are good for leaping. A grasshopper can jump more than three feet (one meter).

Its legs and feet have spikes, which it uses to defend itself against enemies.

Grasshoppers have very keen eyesight and hearing.

Grasshoppers' colors help them blend in with their background.

Name That Tune
Grasshoppers are good fiddle players. They make music the same way a violin produces sound. The grasshopper's leg is its bow, and the tough wing vein is the string. Crickets are also known for their musical ability. They use their wings to make sound. One wing has a thick vein with bumps on it. This is called the file. The cricket rubs the file over a rough ridge on the other wing to make cricket music.

Cricket

Grasshopper

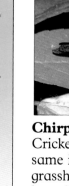

Chirping Cousins
Crickets belong to the same insect group as grasshoppers. But crickets have longer antennae and like eating other insects.

Growing Up

Female locusts lay their eggs in the sand. The babies, called nymphs, hatch and dig their way out. When they appear, they are tiny versions of their parents. In order to become fully grown adults, the nymphs molt between three and five times. After each molt, the nymphs are bigger than before. When they molt for the final time, they emerge with full-sized wings.

Nymph crawling out of nest

Adult pulling itself free of skin

Adult resting while its wings harden

The Trouble with Locusts

When heavy rains fall in hot, dry regions, lush plant life begins to grow. With lots of food, large numbers of grasshoppers get together to mate. After mating, they eat all the plant life around them and grow much larger. In search of more food, they take to the air in huge swarms, devouring fields of valuable crops.

Locust-infested areas

Extended Family

All of these insects are related to grasshoppers, but they have quite different features – and very clever disguises.

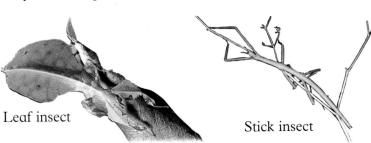

Leaf insect

Stick insect

Praying mantis

BEES

Bumblebee

You have probably heard the buzz of a honeybee as it flies from flower to flower. During the spring and summer months, bees spend their time collecting food. There are thousands of different types of bees, and many of them live alone. But social bees, such as honeybees, live in large nests or hives. They gather nectar to be stored in the hive and turned into honey.

Nest Making
Some bees make their hives out of chewed leaves, mud, and animal hairs.

Working Hives
Hives built by people are specially made to house thousands of bees. Farmers place these hives in their orchards so the bees will pollinate the trees.

As the honeybee moves from flower to flower, bright yellow pollen from each one sticks to its body. The bee carries the pollen back to the hive, where it is turned into food.

This honeybee is feeding on nectar, a sweet liquid found in flowers. It sucks out the nectar with its long, tubelike mouthparts.

Bees help pollinate flowers by carrying pollen from flower to flower.

The bee's sting is in its tail.

African killer bee

Sweet Drink
Worker honeybees look after the young and turn nectar into another sweet liquid, called honey.

A large hive can hold up to 50,000 bees.

Inside the hive, the bees store honey in a comb, which is made up of thousands of little six-sided cells. The bees feed on the honey during the cold winter months.

Orchid bee

Parasitic bee

Shall We Dance?
Worker bees scout for food. When they find a good supply, they do a dance – in a figure-eight pattern – to tell the other bees where the food is. The bees in the hive then know where the food is by the angle of the figure eight and the position of the sun in the sky.

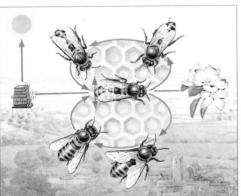

Asian carpenter bee

Queen Bee
Every hive needs a queen. The queen bee mates with a male, called a drone. She then lays all the eggs. New hives are formed in summer when a young queen leads lots of workers out of the old hive to a new one.

Royal Jelly
Royal jelly is actually bee milk. It is filled with good things like sugar, protein, and vitamins. The developing young, or larvae, of the worker bee do not get to eat the royal food – they are fed on pollen and honey. Only the larvae that are destined to become queens eat royal jelly. Because it is so rich in vitamins and proteins, people now use it to make face creams, soap, and vitamin tablets.

Royal-jelly products

125

DRAGONFLIES

Many insects are good fliers, but dragonflies are truly the champions of flight. Millions of years ago, human-sized dragonflies patrolled the skies. Even today's finger-length dragonflies are quite large compared to other insects. Once dragonflies have emerged from their water-based nymph stage, they take to the air, flying at speeds of up to 35 miles (56 kilometers) per hour.

Wing Power
Dragonflies have strong muscles that control the base of the wings. In flight, the wings look like a rapidly changing X shape.

Dancing Dragonflies
To attract a mate, swarms of male dragonflies perform dances in the air.

Mating dragonflies

Baby Dragonflies
When the female is ready to lay her eggs, she dips her abdomen into the water. The eggs sink below the surface.

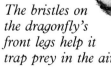

Each pair of clear, veined wings can beat separately. This means that dragonflies can hover.

Dragonflies have excellent eyesight. They have two huge compound eyes. Each eye can have up to 30,000 lenses.

The bristles on the dragonfly's front legs help it trap prey in the air.

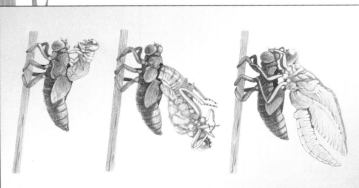

Growing Up
Once the fully developed dragonfly nymph has climbed up out of the water, it begins its last and most spectacular change. The nymph's skin cracks open, and an adult dragonfly pulls itself free.

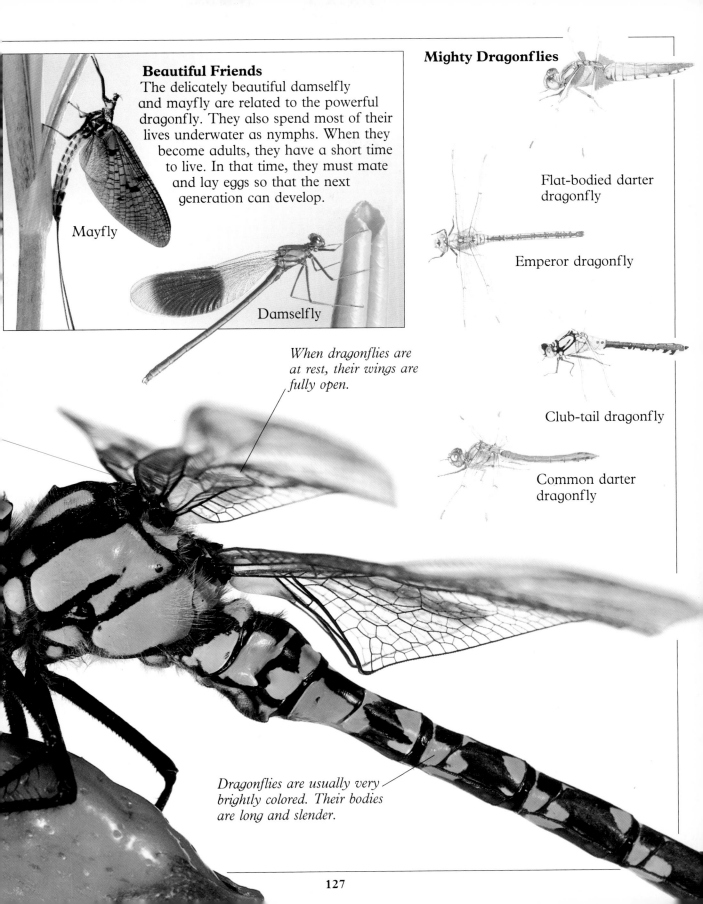

Beautiful Friends

The delicately beautiful damselfly and mayfly are related to the powerful dragonfly. They also spend most of their lives underwater as nymphs. When they become adults, they have a short time to live. In that time, they must mate and lay eggs so that the next generation can develop.

Mayfly

Damselfly

Mighty Dragonflies

Flat-bodied darter dragonfly

Emperor dragonfly

Club-tail dragonfly

Common darter dragonfly

When dragonflies are at rest, their wings are fully open.

Dragonflies are usually very brightly colored. Their bodies are long and slender.

WEB SPIDERS

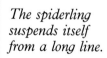

Spiders are afraid of us, so the only signs we tend to see of them are the silken webs they spin to catch their prey. Not all spiders make webs, but the ones that do, like this funnel-web spider, are good at recycling. When its web gets damaged, the spider eats it, digests the silk, then spins a new web.

I'll Eat You Later
If a web spider catches a tasty insect, but is not hungry, it poisons it, but does not kill it. Then it wraps it in silk and keeps it for later.

Up, Up, and Away
When baby spiders – called spiderlings – want to travel long distances, they take to the air. They do not have wings, but they are still able to fly. They produce a piece of silk and use it like a balloon.

Spiders do not have bones. The head and thorax are covered by a hardened shield.

The spiderling suspends itself from a long line.

It makes a loop, which is slowly drawn up by the breeze.

When it's ready to take off, it cuts itself free.

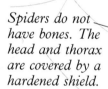

Silk is produced through the spinnerets on the end of the spider's abdomen. Spiders use their legs to pull the silk out.

The saclike abdomen contains the heart, lungs, silk glands, and reproductive parts.

Garden Surprise
Scientists think that insects are attracted to spiders' webs because the webs reflect ultraviolet light. Insects use ultraviolet light to find their way and to find food.

Ready to Attack

This Australian funnel-web spider is one of the world's deadliest. Here it is poised, ready to attack! When spiders catch their prey, usually insects, they use their "fangs" to poison and kill them. The funnel-web spider, like most spiders, uses its strong digestive juices to dissolve the insect's insides so the spider can suck it dry.

Spiders and Webs

The net-casting spider lives in trees, mostly in jungle areas. To catch prey, it spins a sticky net and throws it over passing insects.

The water spider spins a web in the shape of a bell under the water. It fills the bell with a bubble of air and moves in.

Most web spiders have eight simple eyes, called ocelli. But even so, they cannot see very well.

A spider has a pair of graspers, called palps, on each side of its mouthparts. They are used to seize prey.

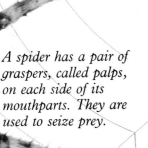

Spiders have eight legs.

Even though spiders do not have ears, they can "listen" to the world around them through their webs. The webs are very sensitive to vibrations in the air.

Making a Web

Making a web takes time and special care. Spiders only spin new webs when the old ones become untidy or damaged.

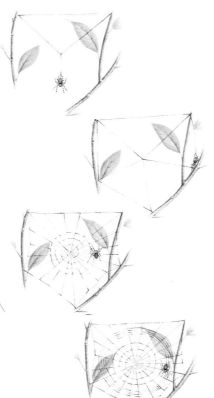

The web of the orb-web spider looks like a target. It takes about an hour to spin a complete web.

The female purse-web spider lives in a silken pouch. When an unsuspecting insect lands on top, the spider bites through the pouch and grabs it.

HUNTING SPIDERS

Many spiders catch their prey without the use of silk. They are called hunting spiders. Some patrol their territory, looking for insects to pounce on. Others crouch in burrows and wait for prey to wander past. Unlike web spiders, hunting spiders have strong jumping legs and keen eyesight so that they can easily spot their victims.

Chilean red-leg spiders are among the largest hunting spiders in the world. They can grow to 5 inches (13 cm) long. They hunt at night and feed on mice and small birds.

A hunting spider has eight simple eyes on the front of its head. It hardly has to move its head to be able to see in all directions.

Hide and Seek
A tree-trunk trapdoor spider is a tricky creature. After it has made a safe, secure burrow with its powerful jaws, it constructs a trapdoor at the entrance. Unseen, it lies in wait behind the trapdoor, ready to leap upon any unsuspecting insects that might crawl by.

The hairs on the spider's body are very sensitive to vibrations made by moving prey.

It has strong legs for digging burrows.

Happy Hunters

Long Jumpers
Hunting spiders need good eyesight because they have to see and chase after their next meal. Many hunting spiders can measure exactly the leap they must make onto their victim as they run along after it.

Full Speed Ahead
Wolf spiders hunt during the day. This Australian wolf spider lurks at the entrance of its silk-lined tunnel ready to race full speed after prey.

Brazilian wandering spider

Crab spider

Raft spider

Wood-louse spider

Take Aim, Fire!
A spitting spider spits when it is hungry. When it spots an insect, it spits a stream of sticky gum from each fang. This glues the insect to the ground until the spider can arrive to eat it.

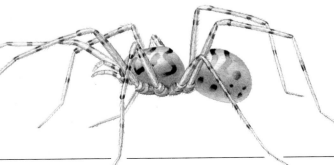

DINOSAURS

The only dinosaur you will ever see is a dead dinosaur! No human being has ever seen a real, live dinosaur because the last of these incredible creatures died about 65 million years ago, long before the first humans appeared. Dinosaur skeletons and lifelike models can help us imagine what these fantastic animals were like. We know that dinosaurs hatched out of eggs and grew up in just a few years. Scientists believe that some lived as we do, in families where the adults took care of their young. Some dinosaurs had large brains and were smart enough to hunt in packs. Others had tiny brains and were not very intelligent.

Dinosaurs were the biggest land animals of all time and different from anything that is alive today.

Tyrannosaurus head

Maiasaura growing up

Camarasaurus skull

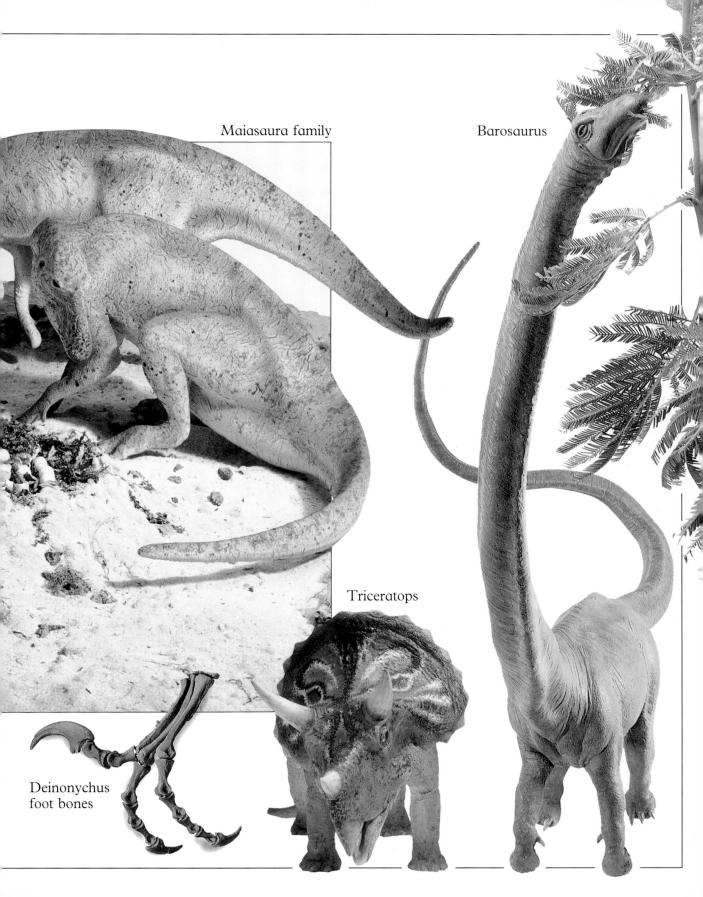

Maiasaura family

Barosaurus

Triceratops

Deinonychus
foot bones

DIFFERENT DINOSAURS

Dinosaurs were first named nearly 200 years ago. There has been an explosion of excitement about them ever since. The huge number of different dinosaur species makes the study of dinosaurs fascinating for everyone. Thousands of bones belonging to hundreds of different dinosaurs have now been discovered. Today, dinosaurs are big business, with millions of dinosaur books and toys for sale in stores.

First and Last

Herrerasaurus was one of the first dinosaurs. It lived about 230 million years ago. Tyrannosaurus was one of the last dinosaurs and became extinct about 65 million years ago. Today, we are closer in time to Tyrannosaurus than Tyrannosaurus was to Herrerasaurus!

Dinosaur Ancestor

Scientists think that over millions of years, all the different dinosaurs may have developed from one single reptile, called Lagosuchus. The hips and long legs of this primitive reptile are similar to those of the earliest dinosaurs.

Classic Jurassic

The middle part of the dinosaur age is called the Jurassic period. It started about 200 million years ago and lasted for 60 million years.

The Age of the Dinosaurs

Dinosaurs didn't all live at the same time. When one species died out, another rose up to take its place. The dinosaur age is split into the Triassic, Jurassic, and Cretaceous periods. The first dinosaurs appeared toward the end of the Triassic period.

Plateosaurus

Dilophosaurus

Heterodontosaurus

Ornitholestes

Handy Hips

Looking at the hipbones of dinosaurs helps us split the dinosaurs into two main groups. If the hipbones are arranged like a bird's, the dinosaur is in the "bird-hipped" group. If the hipbones are arranged like a lizard's, the dinosaur is in the "lizard-hipped" group.

Bird Hips

Iguanodon is in the bird-hipped group, with Triceratops, Euoplocephalus, and Stegosaurus.

Stegosaurus

Euoplocephalus

Triceratops

Iguanodon

Lizard Hips

The lizard-hipped group includes two-legged meat-eaters, such as Tyrannosaurus rex. Plant-eating sauropods, such as Diplodocus, were also lizard-hipped.

Diplodocus

Tyrannosaurus

Cretaceous Creatures

The last stage in dinosaur history is called the Cretaceous period. This stage lasted for about 75 million years. The fiercest meat-eaters lived during this time, alongside their heavily armored prey.

Stegosaurus

Ceratosaurus

Tyrannosaurus rex

Saltasaurus

Torosaurus

Compsognathus

Euoplocephalus

135

DIGGING UP A DINOSAUR

Dinosaur fossils are very rare because they are only found buried in certain types of rock. Dinosaurs became extinct at the end of the Cretaceous period, so any rock that formed after this time does not contain any dinosaur fossils. Rocks at the bottom of the sea are usually free of dinosaur fossils because dinosaurs lived on land. The best places to hunt for dinosaur bones are in rock layers that formed at the bottom of swamps, lakes, or rivers.

This skeleton has been very well preserved. Most of the skeleton is still buried under the ground.

The rock around the fossil has been slowly and carefully chipped away.

Drawing fossils is an important job on a dinosaur dig.

This Tyrannosaurus skeleton is 40 feet (12 meters) long. It is lying curled up on its left side.

Old Bones, New Technology
Modern technology, like this ground-penetrating radar, can be used to find bones below the rock surface. In the 1980s, a dinosaur named Seismosaurus was found in this way in New Mexico.

Dinosaur Trap
The Dinosaur National Monument in Utah is one of the world's richest dinosaur treasure troves. It was once a sandy riverbed, which trapped many dying dinosaurs and preserved their bones.

Handle with Care
Although dinosaurs were strong when they were alive, their fossils are very fragile now. Digging up a dinosaur is a slow and very delicate process.

1. The thick layers of rock above the dinosaur are often cleared away with mechanical diggers and sometimes blasted away with explosives.

2. The last 8 inches (20 centimeters) or so of rock above the dinosaur are removed very carefully with tools.

Fossil Collection
Dinosaur bones are not the only kind of fossils found at dinosaur sites.

Footprints   Skin prints   Stomach stones

Eggs   Droppings   Teeth

3. The exact positions of the bones are then mapped out and recorded with drawings and photographs.

4. The fragile bones must be covered with plaster for protection before they can be moved away from the site.

5. At last, the strengthened bones are loaded onto a truck and driven to a laboratory for further study.

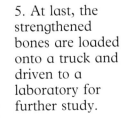

REBUILDING A DINOSAUR

Thick bones in the neck held up the heavy weight of Tyrannosaurus's head.

Jigsaw puzzles can be tricky, especially if some pieces are missing and pieces from other puzzles are mixed in. People who rebuild dinosaur skeletons are faced with a similar puzzle. Fossil bones must be put together in the right order to form a skeleton. Bones belonging to different dinosaurs may have found their way into the collection by mistake. Missing bones are modeled by looking at the bones of similar dinosaurs. Broken bones must be cleaned and mended before they can be assembled.

Back at the Lab

Dinosaur bones are taken from the fossil site to a laboratory. Experts clean and mend them, ready for rebuilding into a skeleton. Some fossils may be prepared only for study, and never displayed in museums.

Bellyache

Early this century, scientists argued over how to rebuild the skeleton of Diplodocus. Some believed Diplodocus was built like a lizard, with long legs sprawling out from its body. This was an impossible pose. Diplodocus would have needed to drag its belly through a trench to walk!

1. First, hard, protective plaster must be removed from the bone with special tools.

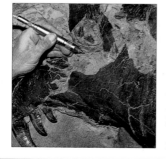

2. Next, the bones must be carefully cleaned. This stage can take a very long time.

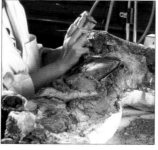

3. The bones are then treated with resin to make them stronger. They must also be repaired, if necessary.

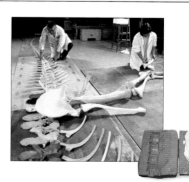

4. Finally, copies are made of the bones using a light material, such as fiberglass. The copies, called casts, can be put together to make a museum display.

Super Skull
Some fossils are almost perfectly preserved. After millions of years buried under rock, this Tyrannosaurus skull did not need much patching up in the fossil laboratory!

Model Making
This model of a Maiasaura nest was sculpted by a scientific artist.

1. The basic Maiasaura shapes are made up with wood and wire.

Building Tyrannosaurus
This skeleton of Tyrannosaurus rex has been on display in an American museum since 1915. Today, scientists think that a big mistake was made in the way the skeleton was put up. They now think Tyrannosaurus held its tail up high above the ground.

A strong rib cage protected its heart, liver, stomach, and lungs.

Huge hips supported the thigh, shin, and foot bones.

Strips of steel run underneath each bone, fitting snugly into the bone's shape.

The metal framework supporting the skeleton is called an armature.

A strong metal pillar holds up the metal strips, supporting the full weight of the skeleton.

2. The nest is built up around the models. The wire skeletons are covered with paper and clay.

3. The clay models are carefully painted a sandy color. In life, Maiasaura was probably well camouflaged.

The tailbones were an extension of Tyrannosaurus's spine.

LOOKING CLOSER

Staring up at the towering skeletons of dinosaurs can make you feel tiny. But bones alone do not give the whole picture of a living, breathing dinosaur. What color was the dinosaur when it was alive? How strong were its muscles, and how big were its heart and lungs? Looking closely at animals alive today gives us useful clues about features, such as the skin color of dinosaurs. We can even imagine what their insides may have looked like!

Brachiosaurus's heart nestled underneath its lungs. It must have had very high blood pressure for the heart to be able to pump blood all around its body.

Weight Watcher
Barosaurus was a very heavy sauropod, like Brachiosaurus. "Sauropod" means "reptile foot." Both dinosaurs had hollow bones in their spines. This was a useful weight-saving device.

Brachiosaurus finished digesting its food in the longest part of its gut, called the small intestine.

Two massive lungs fitted snugly inside its rib cage.

Tough food was broken down inside the thick walls of the stomach.

Moisture was taken out of digested plants in this part of the intestine.

Powerful tail muscles made it possible to swing this bulky tail.

Sturdy leg bones supported the 77-ton bulk of Brachiosaurus.

Strong neck muscles held up its long neck.

Brachiosaurus had nostrils on top of its head.

Skin Deep

Some dinosaurs left behind perfect skin prints. Dinosaur skin was thick, tough, and covered with bumpy scales. The scales did not overlap each other, but were set deep in the skin.

Cunning Colors

We can't tell the color of a dinosaur by looking at fossils. But we can follow clues given by the colors of modern animals.

Chameleons hide from their enemies among the leaves. Hypsilophodon may have been brown to blend in with its dry surroundings.

This brightly colored cockatoo can recognize others of its own kind. Corythosaurus may have had a striking crest for the same reason.

Hadrosaur skin

Scolosaurus skin

The bright stripes of this poisonous snake warn off predators. Deinonychus was a small hunter and may also have been covered with bright warning stripes.

Sauropod skin

Skin and Bones

The big model of Brachiosaurus was made by building artificial muscles onto the plastic skeleton. Then a layer of skin, made from resin, was wrapped over the model.

BIG VEGETARIANS

Plant-eating dinosaurs were some of the biggest animals that ever roamed the Earth. Imagine stepping into the footprint of a Barosaurus – it would be big enough to have a bath in! The big vegetarian dinosaurs called sauropods were peaceful creatures. They grew to their huge sizes on a diet of plants alone.

Gentle Giants
Sauropods had small heads, bulky bodies, and long necks and tails.

Brachiosaurus

A long tail could be useful for whacking an enemy, but most of the time it was used to help Barosaurus balance.

Camarasaurus

Mamenchisaurus

Apatosaurus

Teeth

Stones

Weedy Teeth
Many big vegetarian sauropods had weak teeth. This meant they did not chew plants, but often swallowed them whole. Gulping down stones may have helped grind up food in their stomachs. These stones, called gastroliths, churned tough plants into a pulp.

Its feet were flat and padded, just like an elephant's.

So Long?

The reason why sauropods had such long necks isn't certain. There are three main theories.

1. While Barosaurus stayed in one spot, its neck stretched out for food on land or in the water.

2. Did Barosaurus use its neck as an underwater snorkel, breathing through the nostrils on top of its head? Probably not, because water pressure on its body would have meant it couldn't breathe at all.

3. Barosaurus fed in the treetops. But if it lifted its neck for too long, the blood would have stopped flowing to its brain, making it faint!

Plant Binge

No plant was safe from a sauropod's mouth unless it stood more than 50 feet (15 meters) above the ground!

Conifers are plants with cones. Barosaurus and many other vegetarian dinosaurs ate conifers.

Ferns varied in height from small to tree-sized plants. No fern was too tall for Barosaurus!

Barosaurus ate cycads. These plants still grow in hot climates.

Huge legs supported the crushing weight of Barosaurus.

FEEDING ON PLANTS

Nearly all the huge treetop-munching dinosaurs died by the end of the Jurassic period, about 140 million years ago. The very tall plants disappeared with them. Smaller dinosaurs then arrived on the scene. Iguanodon was well adapted to chewing and chomping. Its jaws were packed with rows of ridged teeth. These grinders pounded away at the leaves Iguanodon nipped off with a sharp, beaky snout.

Chewy Chops
Your cheeks keep food inside your mouth as you chew. Unlike the big sauropods, Iguanodon had cheeks to hold in plant food while it chewed with its teeth.

Big, fleshy cheeks kept food in Iguanodon's mouth as it ate.

A sharp beak nipped off the leaves, and strong teeth at the back of its mouth chewed them into a pulp.

Iguanodon ate plants but not meat, as did all the bird-hipped dinosaurs.

Iguanodon browsed on ferns and horsetails.

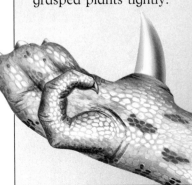

Snatch a Snack
Iguanodon had a sharp thumb claw, which may have ripped down tough leaves. A bent finger grasped plants tightly.

Vegetarian Mouths

The mouths of these dinosaurs were well adapted for eating plants. Heterodontosaurus had three types of teeth for cutting, puncturing, and grinding. Edmontosaurus had a broad snout for gathering up big leafy mouthfuls. Hypsilophodon had a bony beak, with short teeth farther back in its mouth for chopping up its food.

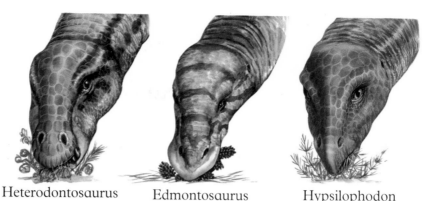

Heterodontosaurus   Edmontosaurus   Hypsilophodon

Jawbone of Edmontosaurus

Rows of teeth for grinding down plants

Beak Mouth

Just like a turtle, Iguanodon had a beak, which it used for nipping plants. This horny beak never stopped growing, but was ground down by a constant diet of tough leaves. Behind the beak was a solid set of teeth, grinding up and down and from side to side.

Fighting Back

Dinosaurs were the biggest plant-eaters of all time. Plants had to find ways of fighting back to survive. Some were so successful, they outlived the dinosaurs and are still around.

The spikes of a monkey-puzzle tree kept away most dinosaurs. Today, no animal will touch them.

Battle of the Flowers

The first flowering plants bloomed about 100 million years ago. They were successful because they could spread their seeds and reproduce more quickly than plant-eaters could gobble them up.

Waxy pine needles taste as bad today as they did in the dinosaur age.

HUNTING IN PACKS

Not all dinosaurs were plant-eaters. Packs of hungry meat-eating dinosaurs roamed all over the land, looking around for their next meal. Deinonychus was a small dinosaur that hunted in a pack. It could outrun its prey and pounce to the kill with frightening accuracy. Dinosaurs much bigger than Deinonychus lived in fear of this clever hunter. Deinonychus was named after its most deadly weapon – its name means "terrible claw."

Running Reptiles
Deinonychus was in a family of meat-eating dinosaurs called the dromaeosaurids, which means "running reptiles."

Compsognathus

Compsognathus fossil

Bavarisaurus

Velociraptor

Lizard Lunch
Many pack-hunting dinosaurs chased other dinosaurs, but the smaller hunters enjoyed a diet of lizards or shrewlike mammals. This fossil of the hen-sized Compsognathus has the bones of its last meal inside its stomach. It was a Bavarisaurus lizard.

The body of Deinonychus was light and speedy. This was ideal for chasing after prey.

Dromaeosaurus

Deadly Claw
Deinonychus had a claw on the tip of each of its toes, but one claw was much bigger than the rest. This huge, sharp talon could swipe around in a semicircle, slashing a deadly wound into the flesh of the dinosaur's prey.

The claw was 5 inches (13 cm) long.

Stenonychosaurus

Brain Box
Each member
of a pack of hunting
dinosaurs had to time its movements
and communicate with the rest of the
group. Deinonychus had a big skull,
which held a large brain, shown here
in red. This means that Deinonychus
would have been intelligent enough
to hunt in a pack.

Why Hunt in Packs?
One wolf cannot attack a deer on its own,
but a pack of wolves can easily pull one down.
The reason for hunting in packs is the same
now as it was at the time of the dinosaurs.
Small, meat-eating dinosaurs joined together
to overpower much bigger dinosaurs.

*Sharp eyesight was
very useful for
spotting prey.*

*Sharp, jagged teeth pointed
backward for a strong,
tearing bite.*

*Often just a quarter of the size
of its prey, Deinonychus was
about 10 feet (3 meters)
long and 6 feet (1.8
meters) tall.*

*This plant-eating Tenontosaurus
bled to death from the gashes
made by the Deinonychus pack.*

147

ENORMOUS MEAT-EATERS

Like tigers today, Tyrannosaurus rex may have hunted alone, terrorizing its prey with surprise attacks. This fierce dinosaur was a carnivore, which means it lived on a diet of meat. The biggest meat-eater ever to walk on this planet, Tyrannosaurus rex was heavier than an elephant and as tall as a two-story building. Its name means "king of the tyrant reptiles."

Hunters need good eyesight and a strong sense of smell. Large parts of Tyrannosaurus's brain controlled its senses of sight and smell.

Tyrannosaurus had a massive skull. It took the shock of crashing into prey at speeds of up to 20 miles (32 kilometers) per hour.

Tyrannosaurus may have charged with open jaws, ready to sink its deadly teeth into its prey. Big chunks of flesh were swallowed whole.

Its short "arms" weren't long enough to reach its mouth, but they may have been used to grip and kill prey.

Duck-Billed Dinner
Tyrannosaurus probably hunted duck-billed dinosaurs, called hadrosaurs. It may have hidden among the trees, waiting for the right moment to charge at a peaceful herd of grazing duckbills.

Life-sized Tooth

A Tyrannosaurus tooth grew up to 7 inches (18 cm) long.

A Tyrannosaurus's tooth was covered with tough enamel.

The sharp edge of the tooth was serrated, like a kitchen knife.

Look how small a human tooth is!

Stretch Those Legs

Tyrannosaurus's short "arms" may also have been used to help the dinosaur get up after a rest on the ground.

1. Small front arms hold its body steady as Tyrannosaurus begins to move.

2. Tyrannosaurus then lifts its head and body backward, stretching out its long back legs.

3. Tyrannosaurus stands up straight. The weight of its tail balances its big head.

Cannibal Dinosaurs

A cannibal is a living thing that eats its own kind. Some meat-eating dinosaurs may have been cannibals. This skeleton of a dinosaur called Coelophysis was found with the tiny bones of a baby Coelophysis inside its stomach.

Meet the Family

The big meat-eating dinosaurs belong to the carnosaur group. This name means "flesh reptile."

Ibertosaurus

Tarbosaurus

Dilophosaurus

Ceratosaurus

Allosaurus

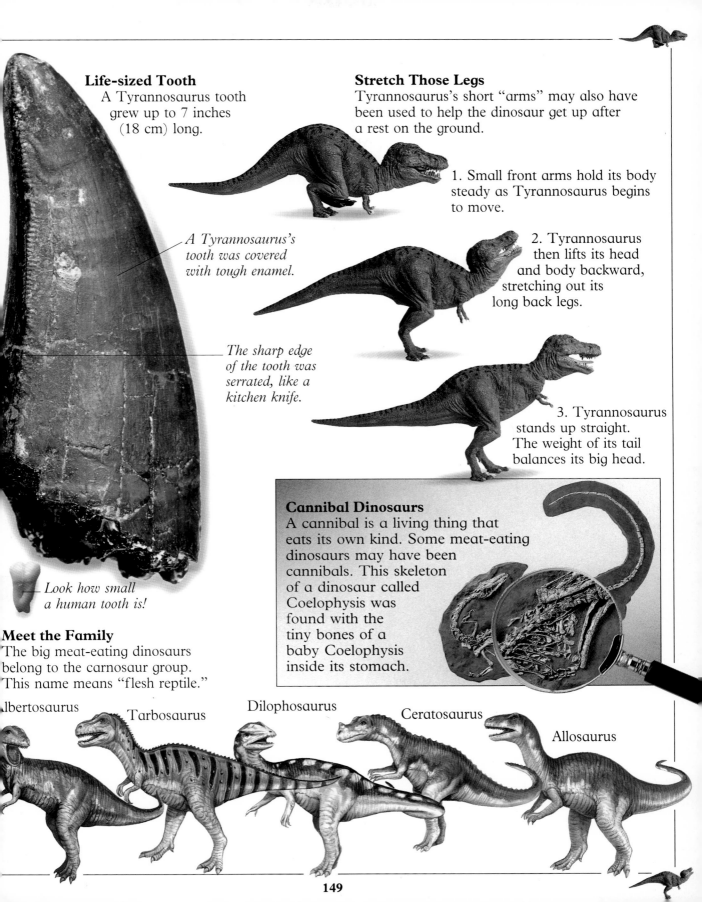

149

DEATH OF THE DINOSAURS

We can tell how old dinosaur fossils are by looking at the age of the rock they were found in. From this, we know that the last of the dinosaurs died out about 65 million years ago. But we can only guess why this happened. Perhaps a huge disaster wiped out all the dinosaurs at once, or they may have become extinct over a very long time.

Rock-solid Proof
The rock on the Earth's surface is formed in layers: the deeper the layer, the older the rock. No dinosaur fossil has been found above the layer that formed 65 million years ago. This proves the last dinosaurs died this long ago.

From Outer Space
Some scientists believe that an enormous asteroid crashed into the Earth at the end of the Cretaceous period. The huge impact of the asteroid could have spelled doom for the dinosaurs.

North America

South America

Yucatán

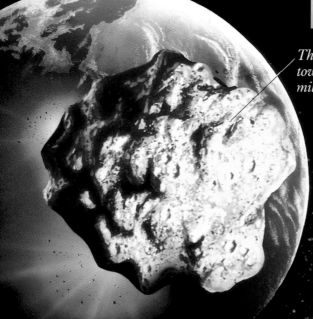

The asteroid may have hurtled toward Earth at about 150,000 miles (241,350 km) per hour.

Slow Change

Dinosaurs may not have disappeared overnight. The world's climate could have cooled over millions of years, slowly killing off the dinosaurs. New plants would have grown in this colder climate. The warm jungles that covered most of the dinosaurs' world would have slowly turned into cooler forests.

What a Hole

A crater 112 miles (180 km) wide has been found in Yucatán in Central America. It may provide evidence for the asteroid theory. This photo shows a similar crater that was made by an asteroid in Arizona.

Wild Guesses

Many theories try to explain the death of the dinosaurs. Some are more unlikely than others! Perhaps hungry mammals ate their eggs.

Or, perhaps the dinosaurs were poisoned by new, nasty plants. But no single theory can explain what really happened.

Shocked quartz

Iridium

Finding Proof

The asteroid crushed rock in the Earth's surface as it hit Yucatán. Shocked quartz has been found at the crater site. Iridium is a rare metal found in asteroids. There are high levels of iridium in the rock that contains the last dinosaur fossils.

The Survivors

Life on Earth did not end with the death of the dinosaurs. Birds, mammals, reptiles, and amphibians are some of the animal groups that survived and are still around today.

Bird

Mammal

Reptile

Amphibian

The disaster would have set off a chain reaction. About 400 trillion tons of rock and dust would have blocked out the sun's light, and the temperature would have dropped to 14°F (-10°C). Acid rain would have fallen on the dinosaurs.

BIRDS

Any animal that grows feathers is a bird. Some are bigger than people; others are almost as small as bees. Birds hatch out of eggs, which are kept in a nest. All birds have feathers and wings, even the ones that do not fly. Because they have no teeth, birds cannot chew. Instead, they grind food in a gizzard. Some food stays in a storage bag, called a crop, so it can be coughed up later for chicks to eat! Birds are like us in the way they breathe oxygen into their lungs. But they are a little like airplanes, too, because their smooth bodies allow them to slip through the air, soaring above snowy mountains, darting through steamy jungles, and skimming over ocean waves. More than 100 billion birds live here with us on Earth.

Blue-tit nest

Parrot feathers

Budgerigar feathers

Owl feather

Green-winged parrot skeleton

Wood duck

Kestrel

Flock of starlings in flight

Blue-tit
egg

Magpie
egg

American
robin egg

Guillemot egg

Gizzard

Crop

Senegal
parrot

FEATHERS AND FLYING

Feathers are made of keratin, like your hair and nails!

Three types of animal can fly – birds, bats, and insects. Birds are the best fliers because they have a coat of feathers. Long, light feathers or small, soft feathers cover almost every part of a bird's body. Usually, only a bird's beak, eyes, legs, and feet are bare. Feathers do much more than cover a bird's naked body: they keep birds warm and dry and enable them to stay up in the air.

Ducks fly in a straight line.

Flapping   *Resting*   *Scarlet tanagers bob up and down as they flap, then rest, then flap again.*

Flight Patterns

Flying is hard work. Not all birds are big and strong enough to flap their wings all the time. Small birds save energy by resting between short bursts of flapping.

Wind Power

If you blow hard across the top of a piece of paper, the air on top moves faster than the air underneath. This difference in airspeed creates lift, and the paper rises. In a similar way, the air moving over the top of a bird's wing moves faster than the air going underneath it. This lifts the bird up in the air.

Waterproof body feathers keep the bird dry.

Underneath its body feathers, the bird wears a warm vest of fluffy down feathers.

This rosella has about 4,000 feathers.

Flapping Flight

As this pigeon flaps its wings down, its feathers close so that they can push against the air. On the way up, the feathers open to let the air slip through.

Adult birds lose old feathers and grow new ones. This is called molting.

Primary feathers can twist like propellers to power the bird through the air.

These secondary flight feathers help lift the bird up in the air.

Vane

Barbule

Hooked Up

Each flight feather has more than one million tiny hooks on it, called barbules. These barbules hook around each other and hold the feather in shape, even in very windy weather.

Barb

Strong shaft

Tail feathers are used for steering and stopping.

Wash and Brush-Up

Birds keep their feathers clean and tidy by preening. They nibble each feather to zip the barbules back together and to get rid of insects. Most birds also waterproof their feathers by rubbing oil into them. This oil comes from a preen gland, which is just above the tail.

Quill

SETTING UP HOME

A nest is a cradle in which eggs and baby birds are kept safe from enemies, such as snakes and rats. Nests can be holes in trees, mounds of earth, or piles of branches. Greenfinches tuck their cup-shaped nests into bushes where they cannot be seen, whereas eagles' lofty nests are easy to see but hard to get at. Each species tries to give its chicks the best chance of survival.

Courting
Before starting a family, male birds have to attract a mate. Peacocks do this by showing off!

Nesting Material
Birds, just like people, build their homes out of all sorts of things. Most nests are made with twigs and leaves, but a few use much stranger ingredients, such as string.

Cattle hair

Seeds

String

Silver foil

Many songbirds glue their nests together with sticky cobwebs!

Birds may make thousands of trips to collect all their nesting material.

Building a Nursery
In many species of birds, both parents build the nest. These long-legged great blue herons are trampling on twigs to make a huge, cup-shaped nest.

Spot the Eggs
Ringed plovers don't build nests. They lay their pebble-like eggs on the beach.

No Teacher Needed

Weavers are brilliant builders, but they don't take lessons or copy other birds. They just know how to weave their nests. This "knowing without learning" is called instinct.

Cup-shaped nests have walls to keep eggs from rolling out.

Baby birds don't need pillows – they have soft feathers to lie on.

Knitted Nest

The weaver uses his beak and feet to tie the grass into knots!

Grass is sewn into the nest to form a ring.

After weaving the walls, roof, and door, he hangs upside down from his home and invites a female to move in.

Nests in trees and bushes are kept dry by the leaves – they form little umbrellas!

Lichen is used to camouflage the nest.

Dried moss helps keep the eggs warm.

This cup-nest was built by squashing! The greenfinch pressed the material into place with its breast as it spun around in a circle.

Burrowing Bee-eater

Carmine bee-eaters nest underground! The male chooses a sandy riverbank and pecks at the earth. When the dent is big enough to cling to, he starts to dig with his beak and feet. The female moves in when all the work is done!

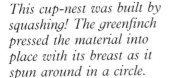

FAMILY LIFE

Family life for most birds is brief, but busy. After the female bird has mated, she lays her eggs, usually in a nest. The baby birds, or chicks, do not grow inside her because this would make her far too heavy to fly. When they hatch, the chicks eat a lot, grow bigger and bigger, and then, as soon as they can fly, they leave home forever.

Baby-sitting
Eggs must be incubated, or kept warm, otherwise the baby birds inside will die. The parents do this by sitting on them.

What's Inside an Egg?
All bird eggs have the same things inside them: a growing baby bird, which feeds on a yellow yolk, and a watery egg white, which cushions the chick from knocks. Woodpeckers and warblers are ready to hatch in 11 days, but royal albatrosses take more than 11 weeks!

Ostrich egg (actual size)

Hummingbird egg (actual size)

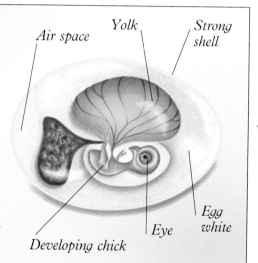

Air space

Yolk

Strong shell

Developing chick

Eye

Egg white

Peck, Peck, Peck . . .
Getting out of an egg is called hatching. The chick taps at the tough shell until it is free. This can take hours or even days!

The blunt end of the egg is pecked off by the chick with a horny spike, called an egg tooth.

First crack

The chick cheeps to tell its parents it is hatching.

Empty shell

Nearly Grown Up
After baby birds have left home, they are called fledglings. Many are eaten by cats and hawks, but some from each nest survive and have families of their own.

Feed Me!
Nestlings are helpless when they hatch: they are blind and naked. All they can do is eat and grow.

The parents push food into their chicks' bright, begging beaks.

Fast Food
Blue tits bring back more than 1,000 juicy caterpillars and aphids to the nest each day for their hungry chicks.

Caterpillar

Ducks' eyes are open when they are born.

The wet, sticky fluff, or down, soon dries out.

After the chick has pushed itself head-first out of the egg, it is very tired.

This egg tooth drops off after a few days.

Follow the Leader
Baby ducks follow the first big, moving thing that they see – usually their mother. This instinct is called imprinting.

Like most waterbirds, just one hour after hatching this duckling can walk, see, swim, and feed itself.

159

THE FROZEN NORTH

The North Pole lies in the middle of a frozen ocean, which is surrounded by a cold, flat, treeless, snowy landscape, called the tundra. Few birds can survive the long, dark winters of the tundra. But in the summer, when the Sun never sets, millions of birds arrive to raise their families. Flowers bloom and the air is full of buzzing insects, but in just a few short weeks, the summer holiday is over and they all fly south again.

The Big Melt
During the summer, the ice melts. But this water can't sink into the frozen ground, so it forms lots of lakes – ideal for ducks!

Bird's-Eye View
Hunting birds have eyes on the front of their heads so that the sight from both eyes overlaps and they can see exactly how far away their victims are. Birds that are eaten by other animals have eyes on the sides of their heads so they can look around for danger.

Blind area

Snowy owl

Ptarmigan

Both eyes can see this dark area.

Only one eye can see this light area.

Tundra Residents

Snowy owl

Snow bunting

Gyrfalcon

Raven

Tough ptarmigans survive the winter by eating dwarf willow twigs.

Birds have a third eyelid that is see-through and helps keep their eyes moist.

Ptarmigans shiver to keep warm.

Snow-colored feathers camouflage ptarmigans.

Causing a Stir

Red phalaropes reach the insects that live at the bottom of lakes by swimming in circles! This stirs up the water and makes the insects float upward. You can try this, too. Put some buttons in a bowl of water, give it a stir, and watch!

Duck Down

Eider ducks pluck soft feathers, called down, from their breast to line their nests. People collect this down to make soft pillows.

Eiderdown

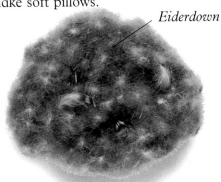

Summer Coat

Willow ptarmigans don't need thick, white coats in the warm summer, so they molt and grow dark red feathers.

Just like big, woolly socks, these feathers keep the ptarmigan's legs, feet, and toes warm.

Willow ptarmigans are about 14 inches (35 cm) tall and live in the Arctic.

Summer Visitors

Ruff

Red phalarope

Red-breasted goose

King eider

COLD FORESTS

One-tenth of the land on Earth is covered in conifers. These tall trees, such as pines and firs, have needles instead of broad leaves and woody cones instead of flowers. The cold forests are home to many birds, from seed-eating siskins to big birds of prey. But conifer forests planted by people, where the trees are all the same type, height, and age, are home to much fewer birds.

The siskin's tail spreads out and acts as a brake.

These feathers are called the alula – they are joined to the bird's thumb.

This Japanese waxwing has a bunch of little feathers, called a crest, on its head.

The feet are stretched out to grab the branch.

Sleeping Safely
When birds bend their legs to sit down, or perch, muscles in their legs make their toes curl. Now the bird can't fall off its perch, even if it falls asleep!

Ant Antics
A Steller's jay gets rid of mites or feather lice by making ants crawl all over its body. The ants get annoyed and squirt out formic acid, which kills the tiny insects!

Angry ants

Waxwings are named after these red dots, which look like drops of wax.

Cone-Opener

Crossbills' beaks can snip open cones a bit like a can opener. Red crossbills eat spruce seeds, but bigger species can open up large pine cones.

Crossed beak

Larch cone

Spruce cone

Pine cone

This greenish brown bird is a female red crossbill – only the males are red!

Inside a Cone

Woody scale

Tasty seed

Starved Out

Every few years, winter is even colder than usual. When this happens, the crossbills and waxwings leave the cold forest to feed in warmer woods.

By opening up its feathers, the siskin can reduce its speed.

Touchdown

Birds have to slow down to land, but, just like planes, if they go too slowly they can stall and fall out of the sky. By sticking out the alula, a bird can change the flow of the wind over its body and avoid crash landings!

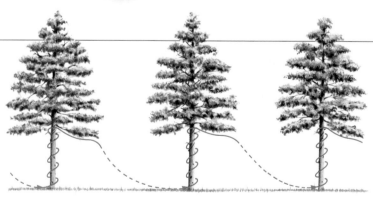

Up and Over

Like a little feathered mouse, the tree creeper creeps up tree trunks looking for spiders and earwigs to eat. It starts at the bottom of the tree, swirls around and around the trunk, walks along the first big branch, and then drops down to the bottom of the next tree.

IN THE WOODS

Europe and large parts of America were once covered by a green carpet of trees. Most of these maple, beech, and oak woods have been cut down to make room for farms and cities. But if you go into the woods that are left, you may still find some of the beautiful birds that live there. Even if you don't spot them through the layers of leaves, you will hear them. Songbirds, such as nightingales, sing flutelike songs both day and night.

First-class Flier
Woodland birds, like the sassy blue tit above, have short wings. This allows them to twist and turn between the trees.

Woodpeckers' feet are designed for climbing trees – two toes point backward.

Pied flycatcher chasing flies

Tough beak

Tawny owls sleep during the day. This one is being woken up by birds who think it is roosting too near their nests.

This greater spotted woodpecker has a hard skull to keep it from getting a headache when it hammers.

Who's There?
Knock, knock – this is the sound a woodpecker makes when it hammers its beak into a tree to dig out insects!

By placing their long, stiff tail feathers against the tree trunk, woodpeckers are able to keep steady while they feed.

Song thrushes open up snails by hitting them against rocks.

European robins sing to claim their territory.

All Fall Down
Most woodland trees are deciduous – they lose all their leaves in winter. Before they drop their leaves, a few, like this rowan, grow juicy berries that many woodland birds love to eat.

Big Baby
Cuckoos are lazy. They lay their eggs in other birds' nests and let them do all the hard work.

A cuckoo has laid its big egg in this dunnock's nest.

As soon as the baby cuckoo hatches, it pushes all the other eggs out of the nest.

The tiny dunnock keeps feeding the cuckoo chick, even when it grows very big!

Crow's nest

Chaffinch eating a caterpillar

This hungry, crafty magpie is waiting for a chance to steal the wood pigeon's eggs.

Wood pigeon

This sparrow hawk is plucking feathers from a hawfinch it has just caught.

Nuthatch

Woodcock

Hawfinch eating seeds

SWAMPS AND MARSHES

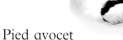
Pied avocet

Half water, half land, swamps and marshes are strange places to live. But to thousands of birds, the shallow water and soggy soil is an ideal home – it is full of food. Birds that live in these tree-filled swamps or murky marshes can either walk through the water or swim. Those that walk are called waders. Wading birds have long, thin legs, sometimes thinner than a pencil!

To keep their eggs and chicks dry, reed warblers build their nests high up in the reeds, well above the water.

Young scarlet ibises have gray feathers. They only grow red feathers when they are several years old.

Black tip to the long, red wing

This is not a knee, but an ankle!

Highly Strung
Foxes, and other egg thieves, can't walk on marshy ground. So, to keep their families safe, reed warblers build their homes in tall marsh plants, called reeds. The nest is tied tightly onto the reeds and won't fall down, even on a windy day.

Scarlet ibises are about 18 inches (46 cm) tall and live in swamps in South America.

Up in the Air
Walking on stilts makes your legs longer. Now you can splash through puddles and your clothes still won't get wet. A little wading bird called the black-winged stilt likes to keep its feathers dry, so it has very, very long legs, which work like stilts.

Fitting the Bill
Storks stab, shoebills dig, flamingos sieve, and spoonbills trap – these amazing beaks, or bills, are all shaped to catch food.

Spoonbill

Saddle-billed stork

Nostril

All birds' legs are covered in small scales.

This long, curved beak can poke deep down into the sticky mud and water at the bottom of the swamp to find frogs and fish.

Shoebill

These big feet keep the ibis from sinking into the mud by spreading its weight over a larger area.

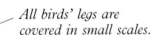

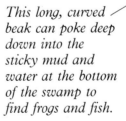

Greater flamingo

A Shady Bird
Sunlight, shining on water, makes it hard to see the fish below. Black herons solve this problem by shading the water with their wings.

LAKES AND RIVERS

Still lakes and flowing rivers provide a well-stocked larder for flocks of ducks, geese, swans, and many more unusual birds. Wherever you live in the world, you can spot these wonderful waterbirds bobbing about on the surface or paddling in the shallows, for they are not shy of people and live on lakes and rivers in cities, too.

Flipper Feet
Water birds have webbed feet. They use these webs of skin like flippers to push through the water.

Splash!
A brilliant blue kingfisher plunges into a river to catch minnows that are half the length of its body. With its wings folded and its eyes and nostrils tightly shut, the kingfisher flies into the water. It has to struggle to free itself from the pull of the water, but it soon succeeds and flies away with its fish.

Treading Water
These western grebes are dancing. When the birds have paired up, they dance across the lake together. Sometimes a male bird tries to steal another's partner!

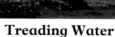

Water skiing   *Gathering speed*   *Liftoff*

Touchdown

Water Runways
Heavy waterbirds can't land or take off on the spot; they need runways, like airplanes!

Finding Food

Different ducks feed in different ways. Some dabble – they dip their beaks into the water to sieve out tiny plants and animals. Others dive several yards under the water to reach weeds, mussels, and insects on the bottom of the lake.

Broad beaks are best for dabbling.

Dabbling ducks never put their whole body under water.

Some diving ducks have notched beaks to hold onto fish.

Shoveler

Teal

Smew

Waterbirds

Black swan

Slavonian grebe

Jacana

Ruddy duck

Mallards are about 16 inches (40 cm) tall and live all over the world.

It is easy to tell male and female mallards apart – males have green heads.

Water drips off a duck's back because its feathers are waterproof.

Male Mallard

Ducks molt their flight feathers all at once, so for a few weeks each year, they can't fly.

Female mallards can quack louder than males.

Female Mallard

Air is trapped between the mallard duck's 12,000 tightly packed feathers to help it float.

These wild mallards may live for 16 years, but most only survive for 2 or 3 years.

Webbed foot

SEABIRDS

Playful puffins and graceful gannets, like all seabirds, feed, preen, and sleep out at sea for most of the year. But each summer they come ashore to lay their eggs. Most nest in huge groups, called colonies. Their families are safer with thousands of pairs of eyes watching for danger. Some seabirds choose to crowd onto steep cliffs, well out of the reach of many egg thieves.

Gull chicks peck at this red spot to make their parents cough up food.

These long wings catch the breeze and lift the gull up in the air like a kite.

Some seabirds spend most of their life in the air, so they don't often use their legs.

Waterproof feathers

Webbed feet

Seabirds do not get tired on long trips because they can glide for hours on end without flapping their wings.

Dive, Dive, Dive!
When a gannet spots a school of fish, it folds up its wings and hurtles into the water. Less than ten seconds later, having swallowed the fish underwater, it is flying again.

A bony flap covers each nostril.

Gannets dive from up to 100 feet (30 m) into the sea and reach speeds of 60 m/h (100 km/h).

The hard skull protects the gannet like a crash helmet!

This spear-shaped bill is ideal for stabbing fish.

Floating on Air
When the wind hits a cliff, it shoots up toward the sky. By stretching out their wings, seabirds, such as puffins, can use this breeze to lift them up to their lofty nests.

Beaky Bird
A puffin's big, bright beak is hinged so that it can snap up fish and still keep a grip on those it has already caught.

Puffin

Herring gull

Gannet

Guillemot

Spinning Eggs
Guillemots do not build nests. Their eggs are pointed at one end, so if they are moved they just roll in a circle and not off the cliff.

Going Up!
A cliff is like a high-rise apartment building. Cormorants nest on the ground floor, in caves near the bottom.

Pebbles That Move
Ringed plovers live on the beach. They are so well camouflaged that they are hard to see.

Cormorant

TROPICAL FORESTS

Giant trees that seem to stretch to the sky form a huge umbrella over the top of a tropical forest. In the shade beneath this green canopy live thousands of weird and wonderful birds. The hot, wet jungles are home to over half of the world's 9,000 species of birds. Noisy parrots gather fruit and nuts, colorful sunbirds sip the juice out of flowers, and harpy eagles swoop down on chattering monkeys.

When a toucan goes to sleep, it rests its big, brightly colored beak on its back.

Toco toucans are about 14 inches (35 cm) tall and live in South America.

This beak is not as heavy as it looks. It is hollow and has thin rods of bone inside for strength and support.

The toucan uses the jagged edge of its beak like a saw to cut through large fruit.

Big Stretch
With its long, clumsy-looking beak, the toucan can reach fruit and berries that are farther away. It picks them up and tosses them into its throat.

With their strong, hooked beaks, parrots can crack open tough nuts.

Parrots come in all colors. But even bright green ones are hard to spot among tropical fruit and flowers.

Red-fronted parrots are about 10 inches (25 cm) tall and live in Africa.

Parrots hold their food with their claws. Some use the right foot and others the left!

This hummingbird's beak is just the right shape to reach to the bottom of the flower.

Fancy Feathers

King of Saxony bird of paradise

Hummingbirds are the only birds that can fly backward!

Hovering Hummingbird
By flapping their wings very fast, hummingbirds can hover near a flower. They then suck out the flower's juice, or nectar.

Magnificent bird of paradise

Cock-a-doodle-doo!
Chickens have been bred from tropical birds, called jungle fowl. These wild birds look and sound very like chickens, but they don't lay as many eggs.

King bird of paradise

White-plumed bird of paradise

GRASSLANDS

Grass grows in the vast spaces between wet forests and dry deserts. These rolling seas of grass have several names: the tropical African plains are known as savannas, and the colder grasslands are called prairies, pampas, or steppes. These green lands are important to people because they make good farmland. To birds, they are just a perfect place to live. There is tall grass to hide in, and seeds, grasshoppers, beetles, and worms waiting to be eaten.

Crowned crane

Two for Dinner
Honeyguides love beeswax, but they can't open a bees' nest. They have to find a big-clawed ratel to rip open the nest for them.

The honeyguide leads the way.

The ratel leaves plenty for the patient bird.

Bees

If an oxpecker sees a lion, it calls very loudly and warns the buffalo of danger.

Oxpeckers have strong, sharp claws for clinging to thick skin.

Oxpeckers
are about 5 inches (12 cm) tall and live in Africa.

The buffalo ignores the oxpecker, unless it pecks inside its ears!

Doctor Oxpecker!
Oxpeckers peck juicy insects, called ticks, out of the skins of buffalo and zebra. This "surgery" helps keep the animals healthy.

Grassland Birds

Western meadowlark

Vulturine guinea fowl

Budgerigar (parakeet)

Gouldian finch

Save the Chicken!

Every year there are fewer and fewer prairie chickens living on the American prairies. This is because the grass is being plowed up and turned into gigantic fields of corn.

Grassland Parrots

Budgerigars are small, green, Australian parrots. In rainy years, there are more "budgies" in Australia than any other species of bird.

Hitching a Ride

Kori bustards kick up thousands of insects as they stride through tall grass. Bee-eaters use the bustard as a perch to catch these swarming flies.

The bustard's back is used as a takeoff and landing strip!

Kori bustards weigh about 50 pounds (23 kilograms)!

White throat

Bee-watching Bird

Dainty, white-throated bee-eaters eat most insects but like bees best. When they spot one, they grab it, kill it, and swallow it whole!

After killing the bee by thumping it against a branch, the bee-eater closes its beak around the bee and squeezes out the stinger.

DRY LANDS

At midday, hot, dusty, dry lands are quiet and appear to be empty. Only at sunrise and sunset, when the air and ground are cooler, do birds come out of the shade to feed and find water. Elf owls are lucky – they eat juicy meat and do not need to drink. Sandgrouse eat dry seeds and have to fly up to 30 miles (50 km) every day to get a drink of water.

Sandgrouse Slurp
Most birds have to tip their heads back to trickle water down their throats. Sandgrouse are unusual – like you, they can slurp up liquids.

Shaggy crest

Roadrunners run, but they can also fly.

In cartoons, roadrunners say "Beep! Beep!" In real life, they rattle their beaks to make a "clack."

Take-away Water
Male sandgrouse sit in puddles! Their belly feathers soak up water like sponges. When they return to their nests, the chicks drink from the soggy coat.

After the cold desert night, roadrunners warm up by standing with their backs to the sun and sunbathing.

Roadrunners are about 12 inches (30 cm) tall and live in Mexico and the southwestern United States.

Roadrunners swallow rattlesnakes that are up to 20 inches (50 cm) long.

Roadrunners can run as fast as an Olympic sprinter!

Like most ground-feeding birds, roadrunners have long legs and toes.

Roadrunners use their tails to balance as they zigzag across the desert.

Building a Sand Castle!

A mallee fowl doesn't build a nest – it makes a massive mound. It piles leaves into a pit and covers them with sand. The rotting leaves give off heat and incubate the eggs.

The chicks will dig their way out.

Sand is scraped away if the eggs overheat.

This little elf owl is sheltering from the scorching sun in a hole made by a gila woodpecker.

Tiny verdins hang upside down from cactus branches to look for insects.

Cactus wrens have scaly legs and tough feathers to protect them from the cactus spines.

Sandy Birds

Pale-colored feathers reflect the Sun's heat and are good desert camouflage.

Gila woodpeckers peck away dead bits of cactus as they hunt for insects.

This western bluebird is keeping cool in a shady bush.

Gambel's quail search under stones for seeds and scorpions.

THE FROZEN SOUTH

Chinstrap Penguin

Antarctica, the land that surrounds the South Pole, is almost empty. The blanket of ice over a mile thick, howling winds, and freezing temperatures keep plants and land-living animals from surviving. But the ocean around this frozen land is full of fish and krill, so the coasts are home to millions of birds. Penguins are the best-known residents of the snowy south, but there are also seabirds such as skuas, petrels, and terns.

Belly Flop

The quickest way to get around on slippery ice is to slide. Penguins can't use a sled like you, so they scoot along on their big bellies!

They push themselves along with their strong feet and wings.

These Adélie penguins have to slip and slide over 60 miles (100 km) from their nesting colonies to the sea.

Egg Thief

Skuas rarely go hungry because there are plenty of penguin eggs and chicks for them to steal. While one skua distracts the parent penguins, the other one grabs a meal.

They can't fly, but penguins are excellent swimmers.

Adélies eat small shrimps, called krill.

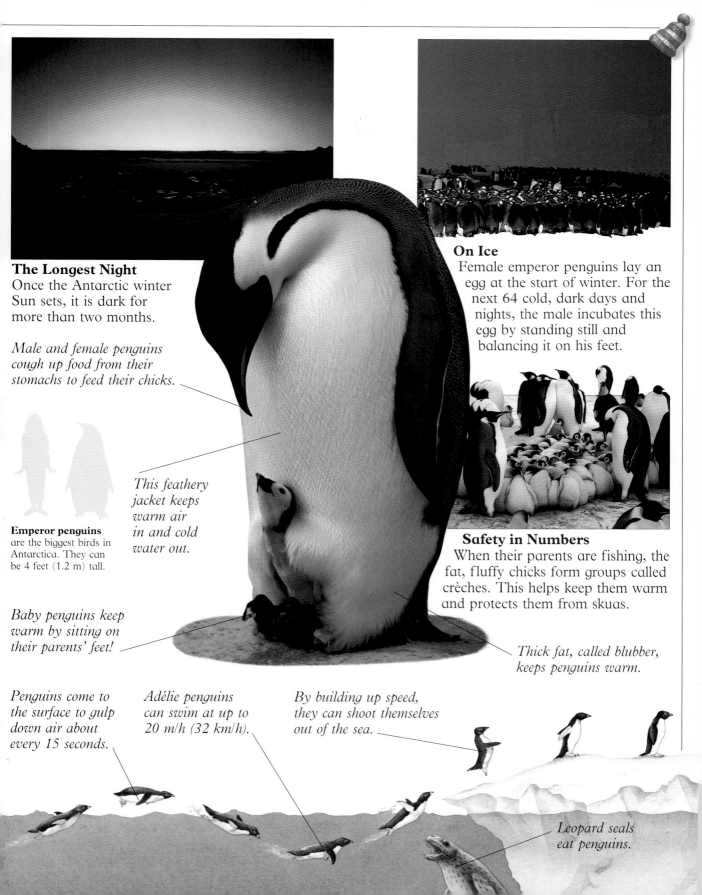

The Longest Night
Once the Antarctic winter Sun sets, it is dark for more than two months.

Male and female penguins cough up food from their stomachs to feed their chicks.

Emperor penguins are the biggest birds in Antarctica. They can be 4 feet (1.2 m) tall.

This feathery jacket keeps warm air in and cold water out.

Baby penguins keep warm by sitting on their parents' feet!

On Ice
Female emperor penguins lay an egg at the start of winter. For the next 64 cold, dark days and nights, the male incubates this egg by standing still and balancing it on his feet.

Safety in Numbers
When their parents are fishing, the fat, fluffy chicks form groups called crèches. This helps keep them warm and protects them from skuas.

Thick fat, called blubber, keeps penguins warm.

Penguins come to the surface to gulp down air about every 15 seconds.

Adélie penguins can swim at up to 20 m/h (32 km/h).

By building up speed, they can shoot themselves out of the sea.

Leopard seals eat penguins.

BIRDS OF PREY

Eagles, hawks, and falcons are birds of prey, or raptors. These strong, fast, fearless birds kill and eat other birds and animals – their prey. Whether they are the size of a sparrow or have a wingspan of nine feet (3 m), like the condor, they all have three things in common: hooked beaks, sharp claws, and "eagle" eyes that can spot rabbits far, far away!

Dressed for Dinner
Vultures poke their heads into dead animals to eat. Their heads are bare, as feathers would get messy.

Skydiving
Peregrines have been timed diving at speeds up to 175 m/h (280 km/h)! At the last moment, they thrust out their feet and stab their victim with their claws.

Wings are swept back and the tail closes like a fan.

The wings are strong enough to lift the falcon into the air even if it is carrying a dead duck.

Its pointed wings help the peregrine falcon fly faster than any other bird.

Pigeon dies and falls to the ground

The tail is used for steering.

The Biggest Nest in the World
Golden eagles' nests, often called eyries, can be over 13 feet (4 meters) wide – bigger than some cars! They do not build a new nest every year but fly back to an old nest and just add a few twigs. Some nests are hundreds of years old.

Peregrine falcons are about 13 inches (32 cm) tall and live all over the world.

Bendy Legs!

African harrier hawks have long legs, which they use to reach eggs, chicks, and bats inside holes in trees. To make it easier to snatch a meal, they can bend their legs backward, forward, and even sideways.

Working Together

Raptors do not hurt humans, and for many hundreds of years, they have been trained to hunt with people. In the Middle Ages, this lanner falcon would have been flown by a squire, a boy who worked for a knight.

The glove keeps the bird's claws from scratching you.

Meat-eating Birds

Birds of prey have excellent color vision.

Nostril

This powerful, hooked beak is used to pull apart animals that are too big to be swallowed whole.

Like all birds of prey, peregrines spend most of the day resting or preening, not hunting.

American kestrel

Goshawk

Many birds of prey have bare legs, but others wear feathered "trousers."

This needle-sharp claw, or talon, is used to grab prey.

All-American Eagle

This U.S. army badge has a bald eagle on it because it is the national bird of the United States.

Harpy eagle

FLIGHTLESS BIRDS

Emu

Kagu

Eye to Eye
Ostriches' round eyes are nearly as big as tennis balls!

Flying is hard work – it takes a lot of energy to flap wings and lift off the ground. For most birds, it is worth the effort because it helps them escape from danger or search for things to eat. But some birds, like kiwis and kagus, live on islands where there are no enemies, and others have found that they can run or swim after food.

Over millions of years, birds such as ostriches, emus, and penguins have gradually lost the ability to fly.

Super Egg
Ostriches lay bigger eggs than any other bird: they are 24 times bigger than a chicken's egg! The shell is so strong that even if you stand on top of an ostrich egg, it will not break.

The shell is about six times as thick as a hen's eggshell.

Its tiny wings are hidden under brown, furlike feathers.

The kiwi's nostrils are on the tip of its long beak.

Sniff, Sniff
The national bird of New Zealand, the kiwi, is one of only a few birds to have a good sense of smell. It sniffs out worms that are in the soil.

Ready, Steady, Go!
Ostriches live on the African grasslands. They cannot fly away from lions and hyenas, but they can run very fast. They can sprint at speeds of 43 m/h (70 km/h) – much faster than a galloping horse.

King penguin

Kakapo

Rhea

Takahe

Brown kiwi

Cassowary

Rheas peck the ground with their big, flat beaks to snap up grass, seeds, and leaves.

These frilly feathers keep the rhea warm at night.

By spreading its wings out like a sail, the rhea can catch the wind and run even faster.

Rheas are 5 feet (1.5 m) tall, weigh more than 45 pounds (20 kg) and live in South America.

Males wave their skinny, bare necks from side to side to attract a female.

Come to Daddy

Rheas are unusual. The males, not the females, incubate the eggs and look after their big babies for up to five months.

Cassowary foot

Ostrich foot

Running birds have massive, muscular legs.

The bones inside this leg are solid. Rheas don't fly, so they do not need lightweight bones.

Rheas look shaggy because the barbules on their feathers do not "zip" together. Flightless birds do not need neat feathers.

Running "Shoes"

The ostrich is the only bird to have two toes; most birds have four toes.

Three front-facing toes

183

NIGHT BIRDS

As the sun sets, most birds settle down to sleep, but some are just waking up. Owls, and other birds that feed and fly in the dark, are called nocturnal birds. They come out at night because there are fewer animals competing for food and many bird-eating eagles are asleep! Owls hoot loudly to one another in the dead of night – they are often heard but seldom seen.

Owls can fly without making a noise because special fringed feathers slow down the air as it rushes over their wings.

Night Fishing
Waders, like this black-crowned night heron, feed in shallow water. So if the tide is out in the middle of the night, that is when they fish.

Slow, Silent Swoop
In the dark, barn owls use their ears, not their eyes, to find their food. They can even hear a tiny mouse chewing a seed.

Many night birds have dark feathers to help them hide in the dark.

Did you spot this nightjar?

Short tail

Spot the Bird
During the day, nightjars sleep on the ground. Their feathers are the color of leaves, so if they stay still, foxes and falcons won't find them.

Wooden Actor
If a frogmouth sees you, it will point its head up at the sky and pretend to be a broken branch. It leaves one eye slightly open, though, just in case you're not fooled.

Face to Face

Long-eared owl

Elf owl

Barn owl

Eagle owl

Owls see ten times better than you in the dark!

Owls cannot move their eyes, but they can turn their necks completely around to look backward.

Like all birds, the ears are small slits hidden under the feathers.

Boobook owls are often called "morepork" owls – this is what they shout!

Boobook owls are about 8 inches (20 cm) tall and live in Australia and New Zealand.

Soft feathers

This curved claw kills rats, mice, lizards, and spiders.

Cough It Up
Owls don't have teeth, so they can't chew their food – mice and birds are swallowed whole! Bones, feathers, and fur cannot be digested, so they are made into pellets and coughed up.

Vole rib

Mouse leg bone

Vole fur

Skull   Hipbones   Jaw   Leg bones   Shoulder blades

MAMMALS

Mammals are amazing animals. Some climb trees. Others race across the ground, burrow, or even fly. Seals and otters are examples of mammals that live in the water. Mammals come in many shapes and sizes – a giraffe is tall, a mouse is small, and a platypus looks like an otter with a duck's beak. So what makes them all mammals? They are hairy, they breathe air into lungs, and they feed their babies milk.

Mammals are warm-blooded, which means their temperature is controlled by their bodies. Most are born live, not hatched from eggs, although platypuses lay eggs. Marsupials are a special kind of mammal because they give birth to babies that are only half-formed. These tiny creatures grow into adults nestled in a warm, safe pouch on their mother's body. Many mammals eat only plants, while others eat meat or insects. You are a mammal, too.

Pale kangaroo mouse

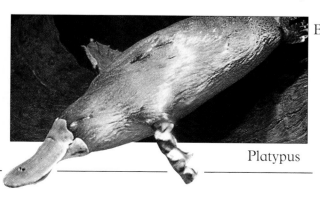

Platypus

Black-tailed jackrabbit

Golden mice

Giant panda

Gelada

186

Snow-leopard cub

Little brown bats

Giraffe

Asian elephant

Tiger

Koala

Giant otter

Royal antelope

Desert hedgehog

MARSUPIALS

A marsupial is an animal that has a pocket, called a pouch, for carrying its babies. Inside this nursery, the baby is safe and has milk to drink. Today, almost all mammals with pouches live in Australia, but 100 million years ago, they lived all over the world. Most marsupials died out when the more modern mammals developed, such as horses, cats, and rats. Marsupials survived in Australia because the "new" mammals could not reach this isolated island. Kangaroos are the most famous marsupials, but there are also marsupial "mice" and "dogs."

Australia

Like deer, kangaroos have long faces to make room for their big, flat, grass-grinding teeth.

Female red kangaroos are three feet (one meter) tall. Males are twice as big. They live in Australia.

Kangaroos can't walk backward!

Female red kangaroos are called blue fliers because they have blue-gray fur and bounce faster than the red males.

The tail helps it balance as it bounds along.

A kangaroo licks a bald spot on its arms to cool down! As the saliva dries, it takes heat away.

Only females have pouches – males don't need them because they don't have babies!

The baby, or joey, hops into the pouch if it sees an eagle or dingo.

Huge leg muscles

Hop to It!
A kangaroo's back legs are so big that it would fall on its face if it ran. So it hops. A red kangaroo can bounce along at 40 miles (65 km) per hour.

Missing Marsupial

The last Tasmanian wolf is thought to have died in a zoo in 1936. It was striped like a tiger and had a thick tail like a kangaroo's. Farmers shot them all because they ate sheep.

Bitty Baby

A newborn wallaby looks like a red bean! It is less than .8 inches (2 cm) long and has no legs, hair, or eyes. Like all marsupials, it continues to grow in a pouch, not inside the mother's body.

Birth

The "bean" squirms through the forest of hair by waving its stumpy arms.

Three minutes later, it reaches the pouch.

It hooks onto a nipple and starts to suck milk.

Acting Star

Opossums are the only living American marsupials. When a Virginia opossum is attacked, it sticks out its tongue, lies very still, and pretends to be dead – it plays possum!

Hold On Tight

This baby koala is too big to fit inside its mother's pouch, so it clings to her fur as she clambers through the eucalyptus leaves!

Mammals with Pouches

Tasmanian devil

Honey possum

Numbat

Ring-tailed rock wallaby

INSECTIVORES

Insectivores are sharp-toothed, long-nosed animals that munch insects, worms, slugs, and snails! Their busy little bodies lose heat easily, so they need to eat a lot. The food they eat produces the energy needed to keep them warm. But how do insectivores survive winters, when there are fewer insects to eat? Shrews search through rotting leaves, and most manage to find enough food. Moles stay underground, and hedgehogs spend cold winters in a deep sleep, called hibernation.

A Bite for Lunch
The water shrew is one of the few poisonous mammals. Its saliva can kill frogs, but not people!

How Hungry?
Imagine having to eat a pile of food that weighs twice as much as you do – shrews have to do this every day!

The tiny eyes are covered by fur. A mole sees poorly – it can just about tell the difference between light and dark.

Little bumps on its tail and its nose help this European mole sense where it is going.

A mole's wide front feet are shaped like spades – ideal for digging.

Molehill

The grass nest is the size of a football.

Worms burrow into the tunnel and are caught by the mole.

Moles turn around by doing forward rolls. If the tunnel is too narrow, they run backward.

A mole eats more than 50 worms a day! Live ones are stored in a larder.

Moles live alone. This worm thief will soon be chased away.

Greedy Guts

Shrews often eat animals that are bigger than themselves. This juicy worm will fill this one's tummy for two or three hours!

Worms in a Week

Streaked tenrecs grow up faster than any other mammals. They stop drinking milk and start to eat worms when they are only six days old.

A hedgehog can stay rolled up for hours.

Insect-Eaters

Golden mole

Roll Up, Roll Up!

Fearless hedgehogs don't run away from danger – they stick out their spines and roll into a ball. No one wants to eat a mouthful of needles!

Head

Desman

Some foxes and badgers have learned to push hedgehogs into puddles to make them unroll!

Star-nosed mole

The hedgehog's skin is larger than its body. When it curls up, it can pull its prickly skin over its head!

Solenodon

Spines are just stiff, hollow hairs.

One-week-old baby shrews hold on to one another so that they don't get lost.

White-tailed shrews

Adult European hedgehogs have more than 5,000 needle-sharp spines.

CATS

A cheetah can accelerate as quickly as a Ferrari c[...]

Cats are carnivores. Most creep up on their prey by sneaking slowly and silently through the undergrowth. Then, suddenly, they will hurl themselves onto their surprised victim. The sharp canine teeth quickly deal the deadly blow. The biggest cat of all, the tiger, can eat 55 pounds (25 kg) of meat in a meal! But this terrifying animal never meets the lion, the king of the cats, because lions live in Africa and tigers in Asia.

Tigers

Bright Eyes
When light shines on a cat's eyes, they glow like the reflectors on the back of a bike. This happens because the light bounces back off a special layer in the cat's eyes. This layer collects light. It helps cats see six times better than you in dim light.

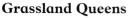

Aerial Ambush
All cats climb trees. This spotty jaguar is waiting to drop down on a passing peccary or tapir. It will even tackle giant alligators.

Grassland Queens
Lions are the cat oddball[ ] because they live in grou[ ] called prides. The male l[ ] is often called the "king o[ ] the jungle," but lions do [ ] actually live in jungles – and the females, or lionesses, are in charge.

Flexible backbone

A cheetah can only sprint for 20 seconds because it soon tires.

Speedy Cat
Cheetahs are the only cats that run down their meals. They can sprint at 60 miles (100 km) per hour and are the fastest mammals.

Play Fight
Baby cats, or kittens, are playful. They chase one another, leap into the air, and chew twitching tails! This is the way they learn to hunt.

Sensitive whiskers

Really Wild Cat
There are more than 300 million pet cats in the world. They are believed to have been bred from wildcats over 3,000 years ago. This Scottish wildcat is much fiercer than a tame tabby!

The rough tongue can rub meat off bones!

Leopards are 2 feet (60 cm) tall. They live in Africa and southern Asia.

Leopards sleep 16 hours a day, usually in short "catnaps."

Big cats roar. They can't purr.

Black leopards, or panthers, can be born to "yellow" parents.

Panthers are spotty, but their spots are hard to see!

Leopards live and hunt on their own. Groups of lions often gang up on them and steal their food.

A cat combs and cleans its fur coat with its tongue.

Soft pads let cats creep quietly.

Put Your Claws In!
Little muscles pull most cats' daggerlike claws into special pockets, to keep them from becoming blunt.

DOGS

Like all carnivores, dogs eat meat. On its own, a dog can only trap animals that are smaller than itself, but 20 African hunting dogs, working together, can easily catch and kill a zebra. Most dogs have learned this lesson and prefer to hunt in family groups, called packs. The 35 types of wild dogs have often been treated as enemies, not loved like pet dogs. Wolves have been wiped out in many places. Foxes survive only because they are smaller and more cunning.

All pet dogs have been bred from gray, or timber, wolves. The first dogs were tamed more than 12,000 years ago!

Hooooowl!

In the dead of night, the wolves in a pack get together, throw back their heads, and howl. This warns other wolves to keep out of their territory.

Gray wolves are about three feet (one meter) tall. They live in the United States, Canada, and northern Asia.

A wolf can hear a watch ticking over 30 feet (10 m) away.

Dogs cool down by panting.

Pointed canine teeth stab the prey. Cheek teeth slice the meat into pieces that are small enough to swallow.

Gray wolves trot more than 35 miles (60 kilometers) a day when they are hunting a moose or an ox.

Pack Property

Female African hunting dogs can have as many as 16 babies. These puppies belong to the whole pack, not just the mother. They are even suckled by other females.

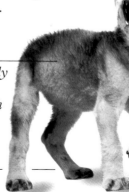

This pup is eight weeks old. It already eats meat, but will not go hunting with the pack until it is six months old.

New Neighbor
Red foxes used to live in woods, but many have moved into cities. They scamper through the streets at night searching for fruit and mice – and garbage cans to raid!

Cleaning Up
Big African dogs, called jackals, love leftover lion food – lions hardly ever finish their dinners! Carnivores that do not kill their own food are called scavengers.

A fox's tail is called a brush.

A gray wolf's thick coat can be any color from white to black!

Win by a Nose
When you smell a flower, you can often tell what sort of flower it is without opening your eyes. Dogs can do much better than this – they can smell who touched the flower the day before!

Dog "Talk"
Every dog has to know its place in the pack – they can't all be the leader! Dogs can't talk, so they use body language instead to let one another know whether they make or take orders.

Bushy tail

Dogs wag their tails when they are happy.

Dogs are marathon runners, not sprinters. A wolf can run at only 28 miles (45 kilometers) per hour – much slower than a lion.

The pack leader holds his tail upright and snarls.

Ankle

This Indian wild dog, called a dhole, does not want to argue, so it rolls on its back.

Dogs walk on their toes.

The claws stay out all the time.

BEARS

Bears are big and usually have thick, shaggy coats. Brown bears are the most common, but giant pandas are more famous. People have argued for years about whether giant pandas are bears or not. Scientists now list the giant panda as a bear – and a very rare one, too! Bears look cuddly, but they are fierce. People have shot so many of these big beasts that today bears survive only in remote areas.

The Big Sleep
Bears that live in cold places spend the winter inside warm caves. The females give birth to their tiny cubs while they are fast asleep.

Polar bear paw print

Honey and Grass for Tea?
Most bears eat all sorts of things – they are omnivores. These are a few of their favorite foods.

Honey

Berries

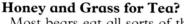

Grass

Masters of Disguise
Polar bears live in the icy Arctic and are the only totally carnivorous bears. Sealskin is their favorite food.

Open Wide!
A grizzly bear has a simple way of fishing: it stands in a river and snaps up fish as they leap out of the water.

The polar bear hides its black nose with its white paw.

It sneaks up on the seal by pretending to be an iceberg!

The cunning bear springs out of the icy water to kill the surprised seal with one swipe of its huge paw.

Half-webbed toes

Ringed seal pup

Paw Prints

Unlike cats and dogs, bears have flat feet. Their heels touch the ground when they walk.

A special pad on the panda's paw is used as a sort of thumb. It is used to grasp bamboo shoots.

All bears have small, round ears.

This big brown bear is over twice the size of a tiger!

A bear's face always looks the same – you can't tell whether it is angry or happy!

Save the Giant Panda!

There are fewer than 1,000 giant pandas left. It is not going to be easy to save them – females are only fertile for three days a year. They also need to eat 45 pounds (20 kilograms) of one special type of bamboo a day.

Many bears have rotten teeth. This is because they love sweet foods – especially honey!

With their big, strong arms, bears can hug a person to death!

Grizzly bears are a type of brown bear. They are called grizzly bears because the tips of their brown hairs are gray, or grizzled.

Grizzly bears stand up to ten feet (three meters) tall. They live in Canada and the United States.

The front paws can be used as clubs to hit other large animals.

197

APES

There are four kinds of apes: chimpanzees, orangutans, gibbons, and gorillas. They all live for many years, have big brains, lack tails, and can walk upright. Apes are the closest relatives of people.

Playful

Begging for food

Orangutan (male)

Gorillas are the biggest and strongest apes, but they are gentle giants. Chattering chimps are clever and cute but much more dangerous – they even kill deer and monkeys to eat! Family life is important to all these intelligent animals. Chimps cuddle and even shake hands when they meet.

Playtime
Baby chimps take a long time to grow up. Their mothers feed them milk for five or six years – so they have plenty of time to play.

Brainy Beast
Chimps are one of the few animals to use tools. They use leaves as sponges! They soften handfuls of leaves by chewing them, then use them to soak up water.

Go Bananas
Gorillas really do eat bananas. They also like nettles, giant celery, and banana leaves.

Apes walk on their knuckles.

Walk like an Ape
All apes can stand on just two feet, but they usually walk on all fours, like this.

Frightened

Angry

Not a Word
The chimp is one of the few mammals that can make faces to show its feelings.

Holding Hands
You can pick things up because you are able to fold your thumb across your hand. Apes and monkeys have these useful, "opposable" thumbs, too.

Like you, apes have sensitive hands.

Male gorillas are about 5.5 feet (1.7 meters) tall, when standing upright, but females are only half this size. Gorillas live in Africa.

An orangutan's big toes can grip things, too!

Gorillas have about the same number of hairs as you. They look more hairy because their hair is long and grows all over their bodies.

Big brain

Gorillas can climb trees, but they spend most of the day lazing on the ground.

Is It a Bird?
Every night, orangs build a cozy nest to sleep in. It takes just five minutes to build a mattress of branches and a blanket of leaves.

Apes see things in color – just like you.

Baby gorillas learn to crawl at ten weeks, climb at five months, and walk at eight months. They may live to be 40.

All apes can sit and stand up straight.

MONKEYS

Monkeys are primates. It is easy to tell them apart from the advanced primates, people and apes, because they have tails. Some, such as mandrills, live on the ground, but most monkeys are light enough to jump or swing through the trees. They always look before they leap, though, because there is danger all around. Large eagles may swoop down, and leopards lurk below. If they lose their footing, monkeys may plunge up to 200 feet (60 meters) to the ground. That's like falling from the 13th floor of a building!

Squirrel monkey

Built to Balance
If you start to lose your balance, you can use your arms to steady yourself. Monkeys use their instead – leaving th arms free for climb

Keep It Clean!
These rhesus monkeys are lining up to have insects and dirt picked out of their fur. They even pick one another's teeth clean! This grooming helps to keep them tidy and also to be good friends.

Groups of monkeys are called troops.

Face to Face

Telltale-Tail
Ring-tailed lemurs are primitive primates and live in troops as monkeys do. They keep together in tall grass by pointing their tails upward.

This lemur is looking for a tail to follow!

Mandrill (male)

Proboscis monkey (ma

Bald uakari

Cotton-top tamarin

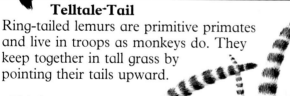

A Gripping Tail
Many South American monkeys have three "arms" – their tails are prehensile, so they can hold on to things. A spider monkey can hang by its strong tail, leaving both hands free for feeding.

Both eyes face forward to spot safe landing places!

All primates, including you, see in color.

Narrow chest

Mane of soft, silky hair

Monkeys are not fussy eaters. They eat fruit, flowers, lizards, butterflies, and even frogs' legs!

Monkeys' legs are shorter than their arms.

This tamarin weighs only 20 ounces (600 grams), so it is light enough to scamper across small branches without breaking them.

People have chopped down so many of the trees golden lion tamarins live in that there are fewer than 200 of these monkeys left. They are in danger of extinction.

Hairy tail

Golden lion tamarins are 9 inches (22 cm) tall. They live in Brazil.

Bathtime
Japanese snow monkeys never get cold feet. They spend much of the winter sitting in pools of hot water that bubble up from beneath the icy ground.

BATS

Bats are furry, flying mammals. Their wings are made of thin, leathery skin, which is stretched across their fingers like material over the spokes of an umbrella. Bats live all over the world, but you will not often see them flying. Most bats hang upside down and sleep during the day. They hunt at night. Small bats eat mice and mosquitoes. Larger, vegetarian bats, often called flying foxes because they have foxlike faces, feed on bananas and nectar.

The ears of this long-eared bat are almost as big as its body. It tucks them under its wings when it sleeps.

Bats hear better than most other mammals.

Acrobatic bats swoop through the sky at up to 35 miles (55 kilometers) per hour – fast enough to escape from owls.

Mouselike, furry body

Wrist

Each wing has the same number of bones as your hand.

Moth

Bats squeak to find one another in the dark. This bat's mouth is full, so it will have to squeak through its nose.

Fruity Bat
Jungles might not exist without bats! Fruit-eating bats pollinate plants. The seeds they spit out or pass in their droppings can grow into trees.

The Bat Cave
Nearly all bats are awake at night – they are nocturnal. Sleepy days are spent inside caves or holes in trees. As the sun sets, millions of bats stream out of their caves to go hunting.

Camping Bats

Imagine having to build a new house every night – tent bats do! These small, white bats nibble through the middle rib of a palm leaf until it droops down to form a tiny tent. The bats hang underneath, out of the wind and rain.

The Smallest Mammal

The hog-nosed bat can fit in the palm of your hand. Its body is 1.2 inches (3 cm) long, and it has a wingspan of just 6 inches (15 cm). It weighs less than one grape.

This is a thumb. Bats use their thumbs as combs to groom their fur and as hooks to hold on to things.

The wing can be used as a scoop to catch flying insects.

Long finger

Long-eared bats have a wingspan of 11 inches (28 cm). They live in northern Europe.

Fangs for Dinner!

Vampire bats love the taste of blood. This one has sliced open the foot of a sleeping chicken with its sharp teeth. It will lap up about one tablespoonful of blood.

Sounds Tasty

Many bats don't use their eyes to see – they use their voices and ears instead! American fishing bats make clicking noises as they fly over ponds. When these sounds bounce back off ripples, they know that a fish is near the surface.

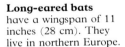

The bat hears the tiny echo and swiftly swoops down to grab the fish.

SMALL RODENTS

A rodent is an animal that gnaws with sharp, chisel-shaped teeth. Most are mouselike and vegetarian: they nibble plant stems, seeds, and roots. Forty percent of all mammals are rodents. They live all over the world – from African jungles, where crested rats climb trees, to scorching deserts, where jerboas hop across the sand. House mice have even hitched rides on boats and lived in huts in Antarctica.

Dormouse

These newborn, rubbery, wriggling mice can only squeak, sleep, and suckle.

Moving House
Every three or four years, thousands of lemmings dash from their overcrowded homes. Many die in the frantic search for new places to live and feed.

A Plague of Rats
Every year, millions of nibbling rats wreck one-fifth of the world's crops!

Many brown rats live in sewers. They use their feet as paddles when they swim and can tread water for three days!

Scaly tail

The greasy fur leaves dirty marks on things it touches.

Flat teeth at the back of a rat's mouth grind up grass and grain.

After walking through dirt, rats walk over food. This is how they spread diseases.

Straw nest

They are 18 days old and ready to leave home.

Their eyes begin to open and fur appears when they are 10 days old.

The Most Common Mammal

House mice are small and cannot defend themselves. The species survives by having lots of babies. A female can give birth when she is six weeks old and have ten litters a year. If all her babies survived, she could have a million descendants when she was one year old!

Out on a Limb

Harvest mice climb up small plants, like monkeys in a tall tree! They hold on tightly with their tails.

Harvest mice weigh only .2 ounces (5 grams) so they rarely break a branch.

Nonstop Teeth

Your adult teeth grow in the spaces left behind by your baby teeth. When they have filled the gap, they stop growing. However, rodents' front teeth never stop growing! To prevent them from getting far too long, they gnaw on things.

Rodents even chew soap.

Growing tooth

Tiny Nibblers

Pale kangaroo mouse

Sagebrush vole

Great jerboa

Indian gerbil

Fat dormouse

Crested rat

LARGE RODENTS

Most rodents are small and look like mice, but some are much bigger. There are two sorts of large rodents – those that look like squirrels, and those, such as porcupines, that look like pigs. All rodents keep their teeth from growing too long by gnawing, but beavers and prairie dogs chew so much that they can transform the countryside.

Chipmunks

Size of a Sheep!
The capybara is the largest rodent in the world. It is a relative of the guinea pig.

The Lion Lost
A porcupine will run backward and stick its quills into an attacker's face!

A porcupine can hear a juicy fruit drop to the ground several yards away.

Hollow, striped quill

The nest, or drey, is the size of a football.

African porcupines chew old bones to keep their teeth sharp.

Angry African porcupines stamp their back feet to rattle their quills. This warns other animals to go away

Giant Nibblers

The sharp quills only stick up when the porcupine is attacked.

European red squirrel

Chinchilla

Rock cavy

Springhaas

Woodchuck

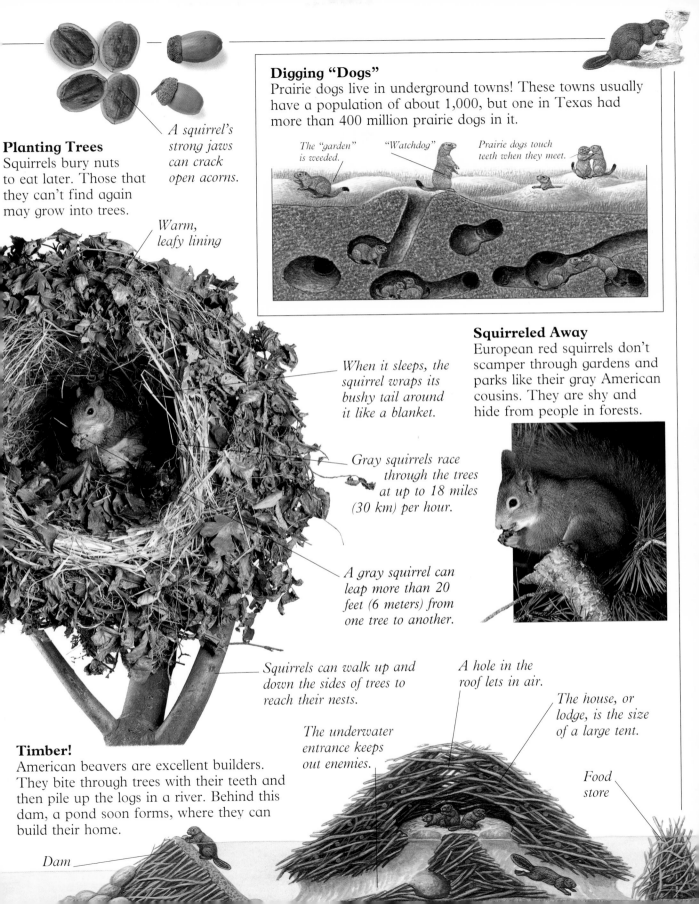

A squirrel's strong jaws can crack open acorns.

Planting Trees
Squirrels bury nuts to eat later. Those that they can't find again may grow into trees.

Digging "Dogs"
Prairie dogs live in underground towns! These towns usually have a population of about 1,000, but one in Texas had more than 400 million prairie dogs in it.

The "garden" is weeded.

"Watchdog"

Prairie dogs touch teeth when they meet.

Warm, leafy lining

When it sleeps, the squirrel wraps its bushy tail around it like a blanket.

Squirreled Away
European red squirrels don't scamper through gardens and parks like their gray American cousins. They are shy and hide from people in forests.

Gray squirrels race through the trees at up to 18 miles (30 km) per hour.

A gray squirrel can leap more than 20 feet (6 meters) from one tree to another.

Squirrels can walk up and down the sides of trees to reach their nests.

A hole in the roof lets in air.

The house, or lodge, is the size of a large tent.

The underwater entrance keeps out enemies.

Food store

Timber!
American beavers are excellent builders. They bite through trees with their teeth and then pile up the logs in a river. Behind this dam, a pond soon forms, where they can build their home.

Dam

THE HORSE FAMILY

Almost all the horses in the world are tame. The only wild species left is the Przewalski's horse. It survives in zoos. The most common wild member of the horse family today is not a horse, but a zebra.

Swift, stripy zebras are herbivores. They graze on grass and chew it with their flat back teeth. Horses, zebras, and asses run on the tips of their toes! Each leg ends in a big toe and a hard "shoe," or hoof.

Przewalski's horse

Desert "Donkeys"
Wild asses are shy and rare. They live in the dry northern African deserts.

A zebra's mane stands up straight.

The ears twist around to listen for danger.

Plains zebras are up to 4 feet (1.2 m) tall at the shoulder. They live in Africa.

A zebra can see in color during the day and as well as an owl at night!

Best Friends
Zebras and wildebeests like to live together. The zebras keep watch, while the wildebeests sniff the air for lions.

Wildebeests eat the short grass left by the zebras.

The hoof is just a large toenail!

Foals can walk when they are a few minutes old.

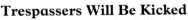

Quick, Run!
Asses, zebras, and horses can run at speeds of up to 30 miles (50 km) per hour.

All four hooves leave the ground when it gallops.

Long legs are best for taking big steps and running fast.

Plains zebras are plump.

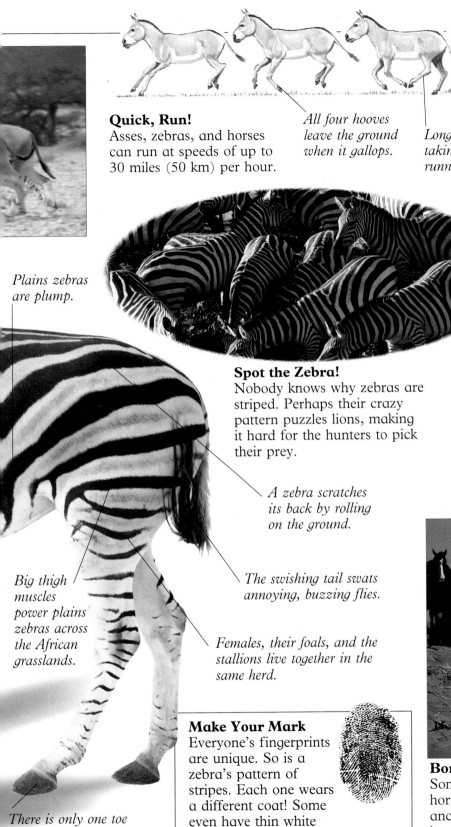

Spot the Zebra!
Nobody knows why zebras are striped. Perhaps their crazy pattern puzzles lions, making it hard for the hunters to pick their prey.

A zebra scratches its back by rolling on the ground.

Trespassers Will Be Kicked
Mountain zebras don't live in herds. Each male, or stallion, "owns" a patch of land, called a territory. He will chase off any male zebra that enters his home.

Big thigh muscles power plains zebras across the African grasslands.

The swishing tail swats annoying, buzzing flies.

Females, their foals, and the stallions live together in the same herd.

Make Your Mark
Everyone's fingerprints are unique. So is a zebra's pattern of stripes. Each one wears a different coat! Some even have thin white stripes on a black background.

There is only one toe inside this hard hoof.

Born Free
Some of the world's 75 million tame horses escaped into the wild. The ancestors of these American "wild" horses, or mustangs, belonged to cowboys and Native Americans.

HIPPOS, PIGS, AND PECCARIES

Hippo is short for hippopotamus, which means "river horse." Hippos are called river horses because they live in rivers and lakes and eat grass. Pigs and peccaries do not live in rivers, but they enjoy wallowing in mud as much as their huge relatives. Although these water-loving mammals are not carnivores, they are all able to protect themselves. Wild pigs can stab and kill tigers with their tusks, peccaries fight jaguars, and a heavy hippo will tussle with a crocodile or smash into a boat!

Very Important Pig
Farmyard pigs have all been bred from wild boars.

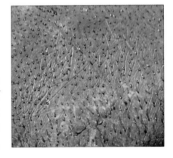

Built-in Suntan Lotion
Hippo skin oozes tiny blobs of pink liquid. This oil keeps their skin from drying out and also protects them from sunburn.

Open Wide
You yawn when you are tired or bored, but male hippos "yawn" when they are angry! Smaller males are frightened off by the big teeth and swim away without starting a fight.

The eyes and nostrils are high up on a hippo's head. This means it can stick just the top of its head out of the water and still see and breathe.

Smooth, almost hairless, skin

The tusks, half as long as a yardstick, are sometimes used to stab crocodiles.

This big male hippo weighs as much as 120 eight-year-old children!

Hippos are 5 feet (1.5 meters) tall. They live in Africa.

Lumbering Lawn Mowers

Every night, hippos leave the water and spend five or six hours grazing. They troop back into the water down well-worn paths long before the scorching Sun rises.

Plucky Peccary

If a mountain lion attacks a group of peccaries, one brave animal runs squealing toward the lion. This peccary dies, but the mothers and babies escape.

Muddy Buddies

Bush pig

Collared peccary

Underwater Ballet

Hippos can hold their breath for more than five minutes. This is plenty of time to dive down and tiptoe gracefully across the bottom of a lake.

There are four toes on each foot.

Hippo

To keep cool and moist, hippos spend 16 hours a day up to their necks in water.

The hippo shuts its ears and nostrils when it is underwater.

Thick skin protects the hippo from snapping crocodiles.

Hippos do eat water lilies, but prefer grass.

Pygmy hippo

THE CAMEL FAMILY

You find camels and their smaller South American relatives, vicuñas and guanacos, on sandy deserts, rocky plains, and bare mountains. They survive in some of the harshest places on Earth. Vicuñas can breathe thin mountain air, and camels can cope with freezing nights and scorching desert days. People have made good use of these animals' amazing survival skills – most are domesticated and work for a living. There are few wild members of the camel family left.

The fat inside a camel's humps provides food to keep it going during hard times.

Domestic Dromedary

Desert peoples could not survive without their one-humped camels. They are ridden, milked, and eaten. Camel skin is made into shoes, hair is woven into clothes, and dry droppings are used as fuel!

Swallow That!

After ten months without a drink, a thirsty camel can gulp down nine big buckets of water in just 15 minutes.

The rumen is the large stomach where chewed food goes first to be partly digested.

Food is brought back up from the rumen to be rechewed.

Second stomach

Third stomach

Twice as Tasty

To get all the goodness out of grass, some mammals, such as camels, deer, and cattle, have more than one stomach and chew their food twice!

The ears and nostrils can be pressed flat to keep out sand.

Two rows of eyelashes keep out sand and stop the eyes from freezing on cold desert nights.

A camel doesn't waste water. Liquid from its runny nose drips down the split lip into its mouth!

Bactrian camels are 7 feet (2.2 meters) tall. They live in the Gobi desert in Asia.

Camels spit at things that annoy them.

Tough lips can grip thorny desert plants.

Camels hardly ever sweat. This saves water.

Spitting Images

Dromedary

Bactrian camel

King of the Castle
While the females graze, the male vicuña stands on a rock. If it spots a mountain lion, it whistles, and the fleecy females flee.

Guanaco

Camels roll from side to side when they walk because they lift both legs on one side at the same time.

Vicuña

The two toes spread out to keep the camel from sinking into soft sand.

Hard-working Mammals
Llamas and alpacas have been bred by people from wild guanacos. Llamas are milked and used to carry heavy loads. Alpacas are kept for their fine wool.

Alpaca

Llama

CATTLE AND ANTELOPES

Cattle, antelopes, and their relatives, goats and sheep, are all bovids. This means that they have horns firmly fixed to the tops of their heads. Horns have a bony core and an outer layer made of the same stuff as your fingernails. Bison graze on grass, but gerenuk antelopes prefer to browse on leaves. Like all bovids, they get goodness out of their poor-quality plant foods by coughing up partly digested food and chewing it a second time. This is called cud chewing.

Gerenuk (male)

All cattle have four stomachs!

These horns are about half as tall as you!

This rare antelope, the Arabian oryx, chews the cud as it walks across the desert.

Arabian oryx

On the March
At the onset of the dry season, huge herds of wildebeest walk almost 1,000 miles (1,600 km) to wetter, greener pastures. When the wet season begins, they wander back. These long, yearly journeys are called migrations.

Like all cattle, bison have split hooves.

It sniffs the air for rain and then walks to where the grass is growing.

Built-in Radiator

The Tibetan yak lives near the top of the world in the Himalayan mountains. It does not get cold, though, because it has its own central heating system – the moss being digested in its stomach is hot and keeps it warm.

This thick winter coat falls off in big clumps during the spring.

A dark coat soaks up the Sun's heat. This helps keep the bison warm in cold weather.

Male bison fight for females by putting their heads together and pushing. The winner is the one who pushes the other backward.

Horns are different from antlers. They never form branches or stop growing, and they are not replaced each year.

Male bison weigh more than a small car!

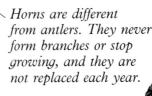

American bison are 6 feet (2 m) tall. They live in Canada and the United States.

Herds of bison spend most of the day eating grass and most of the night chewing!

Heads with Horns

African buffalo (male)

Blackbuck (male)

Wild goat (male)

Bighorn sheep (male)

ELEPHANTS

Matriarch

Elephants have huge ears, long noses, and tusks – and weigh more than six cars. They are the biggest land mammals. Herds of elephants shape the land they live in by treading paths wide enough to stop bush fires, by digging wells in dry riverbeds, by fertilizing the ground with dung, and by trampling grass for zebras to eat. They also open up forests by pushing over trees!

A Family of Females
The leader of a herd of elephants is an old female, called a matriarch. She is followed by all her female relatives and their babies.

Wrinkles trap water and help keep the elephant cool.

Ivory Towers
Many elephants are shot for their valuable ivory tusks. People have burned huge piles of old tusks to show that they want this cruelty to stop.

Elephants never stop growing.

It's All Relative
The elephant's closest relative, the hyrax, looks like a guinea pig! Millions of years ago, hyraxes were huge. All that elephants and hyraxes have in common now are tusks, and nails instead of hooves.

Who's Who?
There are two kinds of elephants.

Tusks are teeth. They grow about 7 inches (17 cm) a year and can be as long as a car!

Ankle

Humped back

Smaller ears

Longer tusks

Bigger, rounder ears

Taller

Elephants eat grass, bark, and leaves for up to 20 hours a day.

This toenail is bigger than your whole hand!

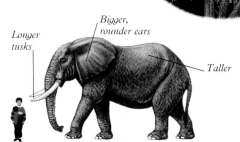

Asian

African

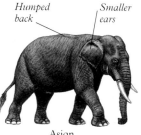

Males leave their families when they are about 14.

Young females act as nannies.

An African elephant's ears are almost as big as sheets for a single bed!

Stay Cool

Elephants have lots of ways of cooling their big bodies. They can wallow in mud or throw water and dust over their skin. Sometimes they flap their ears like giant fans!

The elephant spreads out its ears to make itself look bigger and more dangerous.

Elephants keep in touch by making deep sounds, called "tummy rumbles."

This is one of the first four teeth.

Teeth in the sixth, and last, set are bigger than bricks!

The trunk is formed from the nose and upper lip. It is used for breathing, smelling, touching, and picking things up.

Elephants can live to celebrate their eightieth birthdays!

What's Inside?

Elephants may look like they have flat feet, but they really walk on their tiptoes!

Toe

The heel rests on a fatty cushion.

Always Teething!

Apart from tusks, elephants have only four teeth. These molars are replaced every few years. Bigger teeth appear at the back of the mouth and push out the old, worn teeth – like a conveyor belt of teeth!

THE HUMAN BODY

Your body is one of the most amazing machines in the world. It is made up of thousands of parts all working together. Each group of parts is called a system. The body is so complicated that it is easier to imagine the different systems separately. But your digestive system, nervous system, skeleton, blood or circulatory system, and muscles all work together. You need all of them to stay alive.

When you are born, all you can do is sleep, eat, and cry. It takes time to discover how the parts of your body move and to get them to work together. As you grow, you learn to do more difficult things, like crawling, walking, and riding a bicycle.

There are over 100 muscles in your face.

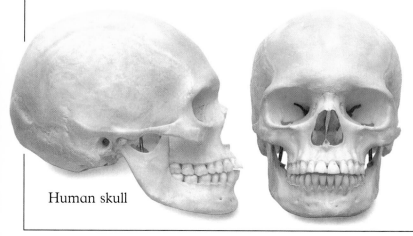

Human skull

Digestive system

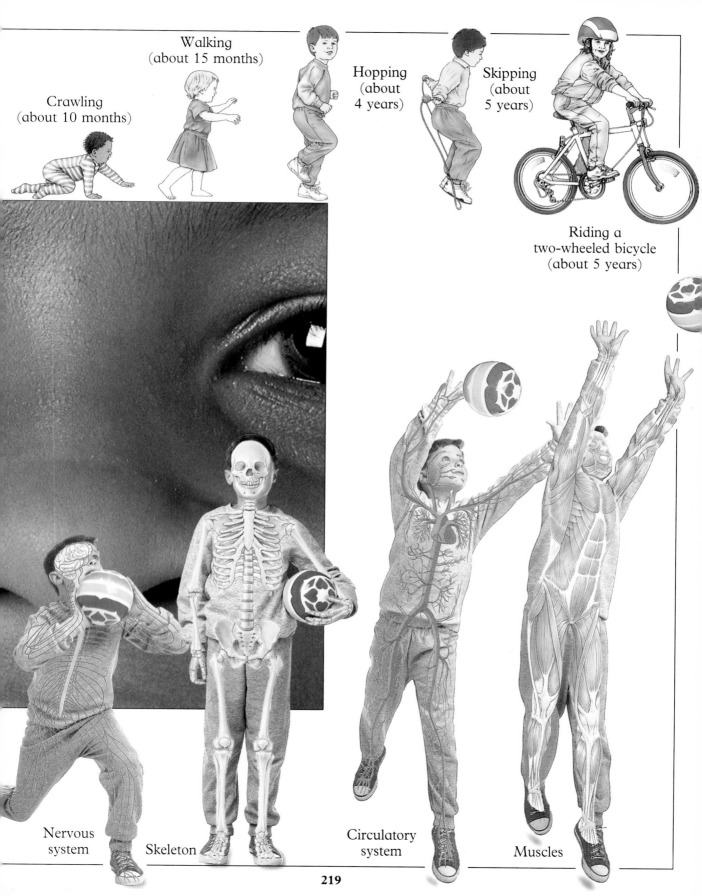

Crawling
(about 10 months)

Walking
(about 15 months)

Hopping
(about
4 years)

Skipping
(about
5 years)

Riding a
two-wheeled bicycle
(about 5 years)

Nervous
system

Skeleton

Circulatory
system

Muscles

HEARING

Hearing is one of your five senses. Your ears are important and delicate: they pick up sounds and send messages to your brain. Sounds travel through the air in waves. Your outer ears, the shell-shaped flaps on the sides of your head, catch sound waves and funnel them inside. These waves hit the eardrum and make it vibrate. The middle ear and the inner ear change the vibrations into electrical signals, which are sorted out and recognized by your brain.

Mother

Kneel

The outer ear collects the sounds and funnels them along the ear canal.

Crash!
The loudness of sound is measured in decibels. Quiet whispering is less than 25 decibels, the clash of cymbals about 90, and a jet plane taking off can measure more than 130 decibels. Noises over 120 decibels can cause pain and may damage your ears.

Hunger

Talking in Signs
If deaf people have never heard speech, they may not have learned to talk. To communicate, many learn a special language, called signing. They also learn to spell using their hands and to lip-read by watching the shape of people's mouths as they speak.

Hope

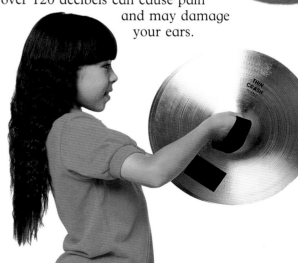

Three tiny, connected bones in the middle ear – the hammer, the anvil, and the stirrup – pass sound vibrations from the eardrum to the cochlea.

The semicircular canals help you balance.

Nerves

Hello . . . Hello!
If you shout in a large, empty space, the sound waves from your voice bounce off the nearest surface, back to your ears. This is called an echo. The farther a sound has to travel before it is reflected, the longer you must wait for the echo.

Nerve cells in the cochlea send messages about sounds to the brain.

The inner ear is made up of the cochlea and the semicircular canals. It is filled with liquid.

The eardrum is a thin sheet of skin that vibrates when sound waves hit it – just like the skin on a drum.

Visible and Invisible Ears
Not all animals hear in the same way as you. Some do not have any ears on the outside of their bodies.

Elephants have huge ear flaps and good hearing.

Birds and lizards have good hearing but no ear flaps on the sides of their heads.

A rabbit can turn its ears to hear sounds all around it.

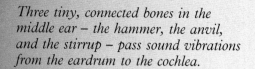

SEEING

Tear-Full
Tears help to keep your eyes moist and clean. When too many are made, they cannot all drain away down your nose, so tears come out of your eyes.

Sight is perhaps the most important of your five senses. Your eyes work together with your brain to help you see. The cornea, at the front of your eye, bends light. This light is then focused by the lens to form an image of what you are looking at on the retina at the back of your eye. But the image is upside down! Nerve cells in the retina send messages to your brain, which sorts the messages out so that you see things the right way up.

Tricky Eyes
Sometimes your eyes play tricks on you, called optical illusions. Try these:

Which of these two girls is taller?

Which red circle is bigger, the left one or the right one?

Which line is longer?

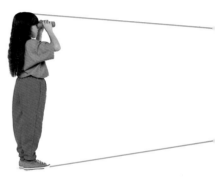

Double Vision?
Because they are set apart, each of your eyes sees a slightly different picture. Your brain puts the two pictures together. This is called stereoscopic or binocular vision, and it means you can judge depth and distances. Try this: close one eye, point to something, and keep your finger still. Now look through the other eye. Are you still pointing to the same place?

Eye Spy
You can have your eyes tested by an optometrist to make sure they are working properly. Glasses or contact lenses will help if you cannot see clearly. Your eyes are very important, so have them checked regularly!

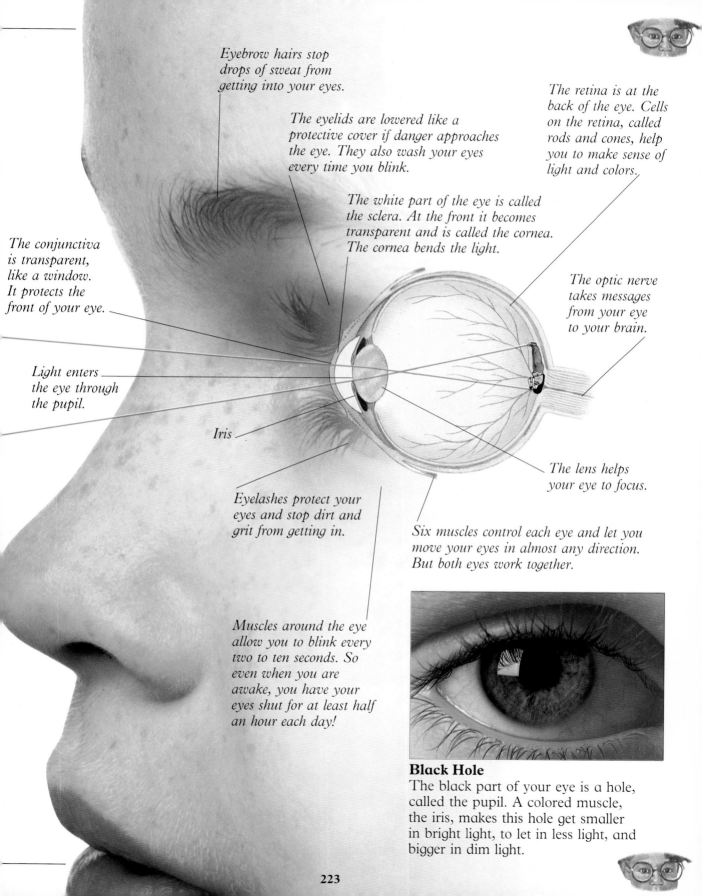

Eyebrow hairs stop drops of sweat from getting into your eyes.

The eyelids are lowered like a protective cover if danger approaches the eye. They also wash your eyes every time you blink.

The retina is at the back of the eye. Cells on the retina, called rods and cones, help you to make sense of light and colors.

The white part of the eye is called the sclera. At the front it becomes transparent and is called the cornea. The cornea bends the light.

The conjunctiva is transparent, like a window. It protects the front of your eye.

The optic nerve takes messages from your eye to your brain.

Light enters the eye through the pupil.

Iris

The lens helps your eye to focus.

Eyelashes protect your eyes and stop dirt and grit from getting in.

Six muscles control each eye and let you move your eyes in almost any direction. But both eyes work together.

Muscles around the eye allow you to blink every two to ten seconds. So even when you are awake, you have your eyes shut for at least half an hour each day!

Black Hole

The black part of your eye is a hole, called the pupil. A colored muscle, the iris, makes this hole get smaller in bright light, to let in less light, and bigger in dim light.

BRAIN

Inside your head, protected by a bony skull, is your brain. It looks crinkled, like a walnut, and is the control center for the whole of your body. Your heart pumps oxygen-filled blood into it through a mass of tubes, called arteries. After only four minutes without oxygen, brain cells will die, and they cannot be replaced. Your brain is divided into two halves, called hemispheres. Each half controls different activities.

Top Gear
To keep from injuring you brain, you should wear a helmet whenever you take part in a sport that might make you bump your hea

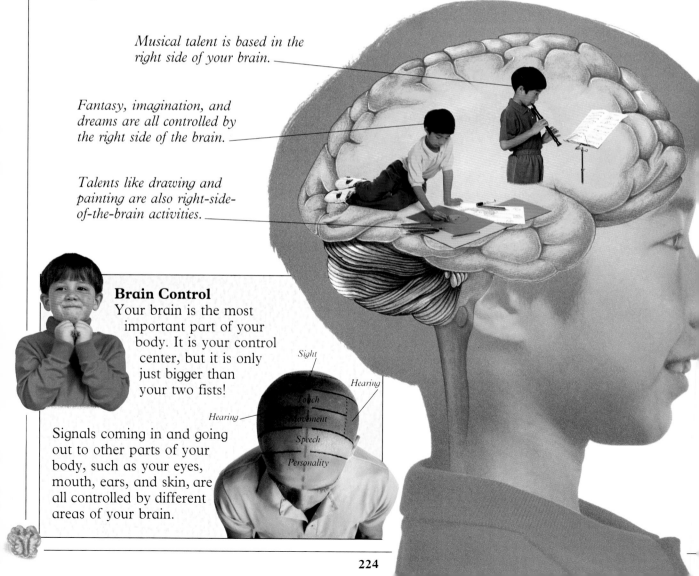

Musical talent is based in the right side of your brain.

Fantasy, imagination, and dreams are all controlled by the right side of the brain.

Talents like drawing and painting are also right-side-of-the-brain activities.

Brain Control
Your brain is the most important part of your body. It is your control center, but it is only just bigger than your two fists!

Signals coming in and going out to other parts of your body, such as your eyes, mouth, ears, and skin, are all controlled by different areas of your brain.

Sight

Hearing

Touch

Hearing

Movement

Speech

Personality

Memory Game

When we are learning, we depend on our memory to help us. Test your memory with this game. Set out a number of objects and look at them for a few moments. Ask a friend to remove an object. Then try to tell what is missing.

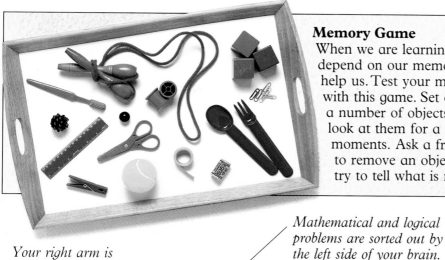

Different Brains

Animals have brains of various shapes and sizes, suited to the things they do.

Fish

Bird

Cat

Human

Mathematical and logical problems are sorted out by the left side of your brain.

Your right arm is controlled by the left side of your brain.

The left side of your brain remembers names, dates, and facts.

The left side of your brain controls language skills. It enables you to speak, read, and write.

The brain stem controls many of your automatic actions, such as your heartbeat and breathing.

Brain Relay

When you want to touch something, a signal is passed from your brain to other parts of your body like a baton in a relay race. It goes first to the spinal nerves and then to the motor nerves, which tell muscles to move.

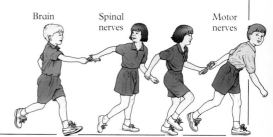

Brain   Spinal nerves   Motor nerves

Left, Right, Left
The ideas in these two large pictures are thought to be true for right-handed people. In left-handed people, they may be reversed.

NERVES

Nerves are like telephone cables carrying information between the brain and all parts of the body. One set of nerves – the sensory nerves – carries signals to your brain from your senses, telling your brain what is happening around you. When your brain has decided what to do, it sends signals along another set of nerves – the motor nerves – to make your muscles work.

Nerve network

Speedy Reflex
If someone claps their hands by your face, your brain thinks something is flying toward your eyes, and they blink. This blink is a reflex action to protect them.

With your eyes covered, can you tell it is a teddy bear?

What's on the "Feely Table"?
The sense of touch is very important. It is one of your five senses. Receptor cells under your skin send messages, through sensory nerves, to your brain about what your fingers are feeling.

A bottle feels hard and smooth.

Reading by Touch
Blind people cannot see to read, so they learn a special alphabet of raised dots, called braille. They feel these dots with their fingertips.

Get the Message?
Messages about how objects feel are sent through the sensory nerves and the spinal nerves to the brain to be understood.

Sensory nerves   Spinal nerves   Brain

If you touch some honey, your fingers will find that it is sticky.

Ice is cold and slippery. When it starts to melt, it feels wet.

Skin Sense

Find out which part of your skin is most sensitive with this test. Close your eyes and ask a friend to touch different parts of your body with the points of two pencils. Try this test on your fingertips and on your knee to see which parts feel two points and which feel only one.

Blindfold

Your fingers can tell if something is soft. It feels good to use towels made of soft material.

Hot things are good to feel when you are cold – but if something is too hot, your nerves will send messages to make your hand move quickly away.

The wood feels hard and rough.

Even when you are blindfolded, you can feel that this smooth, round object is a ball.

The bristles on this brush are sharp and prickly.

Feathers are very light and soft. They can sometimes tickle, especially if you stroke your face gently with them.

Your eyes tell your brain that something may be heavy. Your muscles are then prepared to lift a heavy weight. If you cannot see the weight, you may be surprised when you try to lift it!

SKIN

If you could unwrap your skin, you might be surprised at how much you have – enough to cover a large towel. It goes over all your curves and into every crease of your body. Your skin grows with you, so that when you are an adult, it will cover an area of about 2 square yards (1.7 square meters). It is waterproof and protective and can heal itself if it gets damaged.

New skin grows all the time to replace old skin that rubs off. Most house dust is really old, dead skin!

The thinnest skin is on your eyelids.

Personal Prints
Fingerprints are patterns of lines and swirls on your fingers. Try printing yours using paint or ink! Everyone has different fingerprints, so they are used to identify people. There are three basic patterns.

Arch   Loop   Whorl

Hairs grow on every part of your skin except for your lips, the palms of your hands, and the soles of your feet.

A fingernail takes about six months to grow from base to tip. It grows about 0.02 inches (0.5 mm) a week.

Nails are made of dead cells that contain a protein, called keratin.

The cuticle is the fold of skin overlapping the nail bed from where the nail grows.

The half-moon at the base of your nail looks white because this part is not firmly attached to the skin below.

Shades

All skin contains a coloring substance, called melanin. Dark skin contains more melanin than fair skin. As a protection from burning by the Sun, the skin produces more melanin, which tans the skin a darker color.

The roots of your hair are alive and grow about 0.08 inches (2 mm) a week. When the hairs reach the surface of your skin, they die – so having your hair cut does not hurt!

Growing Old

Your skin is elastic. If you pinch the back of your hand and then release it, the skin goes back to its original place. As people get older, their skin becomes less elastic and gets wrinkled. Some people's hair turns gray because it stops making melanin.

Dead or Alive?

Your skin has two layers. The dead, outer layer, called the epidermis, protects the living dermis underneath.

Hair

A pore is the opening of a sweat gland.

Epidermis

Nerve ending

Dermis

Oil glands keep skin from drying out.

Sweat glands open to let moisture out of your skin to help keep you cool.

Hair follicle

Blood vessels bring food and oxygen to the skin.

Hair Styles

How straight or curly a hair is depends upon the shape of the pocket, or follicle, it grows from. The color of your hair depends upon the amount of black or red melanin there is inside these pockets.

BLOOD

Blood is pumped all around your body by your heart. It travels in long tubes, called blood vessels. Before it begins this journey, it is pushed to your lungs to collect oxygen. Then it returns to your heart to be pumped around your body. Blood also carries nutrients from your food to the cells.

Arteriole

Artery

Aorta
(largest artery)

There . . .
Blood vessels that take blood away from your heart are called arteries.

Power Pump
Your heart is made of strong muscle. In just one minute, it can pump a drop of blood all the way down to your toes and back to your heart again.

The heart is divided into four spaces, called chambers – two at the top and two at the bottom. There is a wall of muscle down the middle.

Blood enters the top right chamber through the vena cava. It passes down to the bottom right chamber. Then it is pushed out to your lungs.

Blood filled with oxygen from the lungs comes into the top left chamber. It passes down to the bottom left chamber. Then it is pumped around your body.

After going around your body, the blood returns to the right side of your heart to start the journey again.

Aorta

Flaps, called valves, stop blood from flowing the wrong way. It is the "lub-dub" sound of these doors closing that you hear when your heart beats.

Venule

Vein

Vena cava
(largest vein)

. . . and Back
Those that take blood back to your heart are called veins.

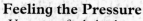

Feeling the Pressure
You can feel the beat of your heart as it pumps blood, at high pressure, through your arteries. This beat is called the pulse. Put the fingers of one hand on the inside of your other wrist, in line with your thumb. You can feel your pulse beating there. You can only hear a heartbeat by using a stethoscope or by putting your ear to a friend's chest.

Heart Work
When you skip, or do any form of exercise, your muscles need extra oxygen and food from your blood. To provide this, your heart has to pump faster.

You have about eight pints (four liters) of blood in your body. An adult has more, and a baby has less.

A drop of blood goes around your body more than a thousand times a day. Every five minutes, all your blood passes through your kidneys to be cleaned.

Super Cells
In one tiny drop of blood, there are red cells, white cells, and platelets, all floating in a liquid, called plasma. The blood in your arteries is carrying oxygen, which makes it a brighter red color than the blood in your veins.

Side view of a red blood cell

Red cells carry oxygen around the body. They are made in the bone marrow.

White blood cells defend the body against germs.

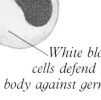
If you cut yourself, platelets rush to the broken blood vessel and stick themselves together to plug the hole.

BREATHING

You must breathe all the time to stay alive. If you try to hold your breath for more than about a minute, your body will force you to start breathing again. The air that you breathe is made up of many different gases mixed together, but your body only needs one of them, oxygen, to keep you alive. If you ran out of oxygen, even for a very short time, you would die. The air you breathe goes into two soft, moist sponges, called lungs. You have one on each side of your chest.

Lung Capacity
Breathe in deeply. Blow into a balloon until you run out of breath. Tie a knot in the balloon. Now you can see just how much air your lungs are able to hold.

Bad Breath!
The air that you breathe can contain gases that are bad for you. This woman is wearing a mask to protect herself from car fumes.

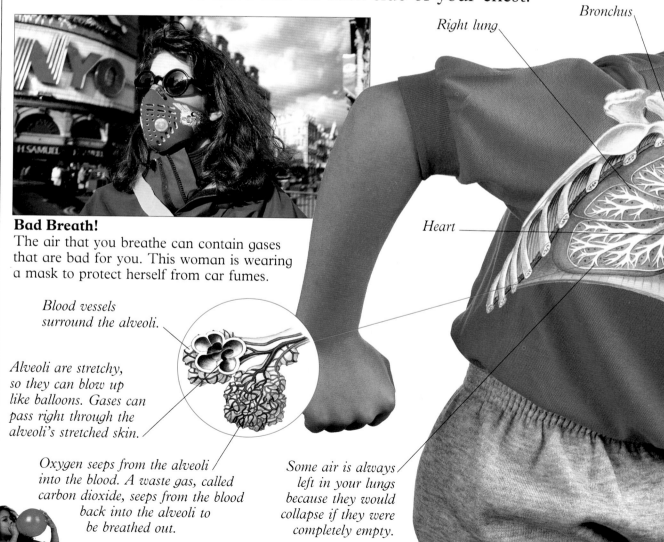

Right lung

Bronchus

Heart

Blood vessels surround the alveoli.

Alveoli are stretchy, so they can blow up like balloons. Gases can pass right through the alveoli's stretched skin.

Oxygen seeps from the alveoli into the blood. A waste gas, called carbon dioxide, seeps from the blood back into the alveoli to be breathed out.

Some air is always left in your lungs because they would collapse if they were completely empty.

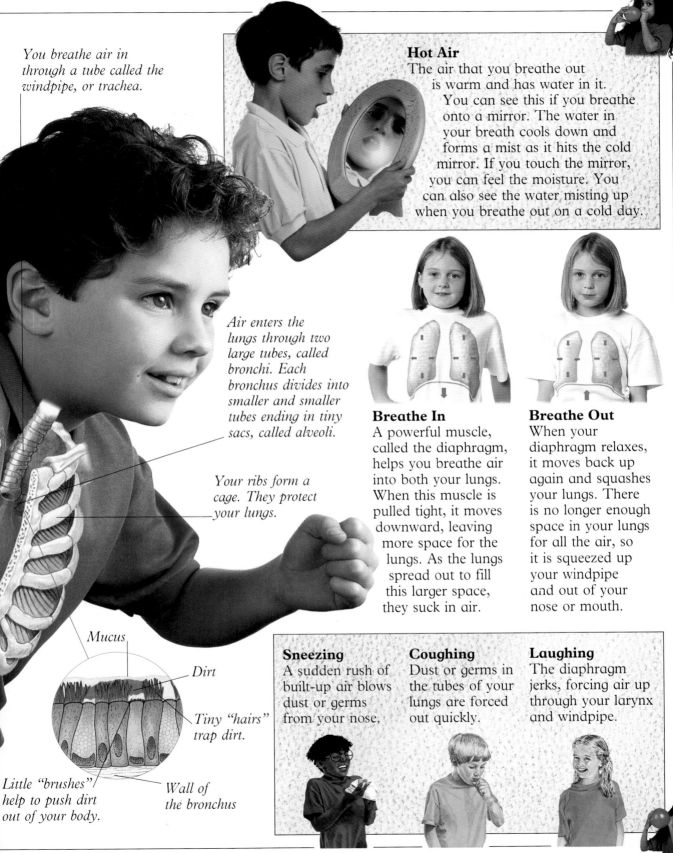

You breathe air in through a tube called the windpipe, or trachea.

Hot Air
The air that you breathe out is warm and has water in it.

You can see this if you breathe onto a mirror. The water in your breath cools down and forms a mist as it hits the cold mirror. If you touch the mirror, you can feel the moisture. You can also see the water misting up when you breathe out on a cold day.

Air enters the lungs through two large tubes, called bronchi. Each bronchus divides into smaller and smaller tubes ending in tiny sacs, called alveoli.

Your ribs form a cage. They protect your lungs.

Breathe In
A powerful muscle, called the diaphragm, helps you breathe air into both your lungs. When this muscle is pulled tight, it moves downward, leaving more space for the lungs. As the lungs spread out to fill this larger space, they suck in air.

Breathe Out
When your diaphragm relaxes, it moves back up again and squashes your lungs. There is no longer enough space in your lungs for all the air, so it is squeezed up your windpipe and out of your nose or mouth.

Mucus

Dirt

Tiny "hairs" trap dirt.

Little "brushes" help to push dirt out of your body.

Wall of the bronchus

Sneezing
A sudden rush of built-up air blows dust or germs from your nose.

Coughing
Dust or germs in the tubes of your lungs are forced out quickly.

Laughing
The diaphragm jerks, forcing air up through your larynx and windpipe.

TASTE AND SMELL

Your senses of taste and smell are very closely linked. They depend on each other. Tiny dimples on your tongue, and hairs at the top of the inside of your nose, detect chemicals that cause tastes and smells. Special sensory cells then send messages through to your brain to be recognized. Your sense of smell is about 20,000 times stronger than your sense of taste! Often what you think you are tasting, you are really just smelling.

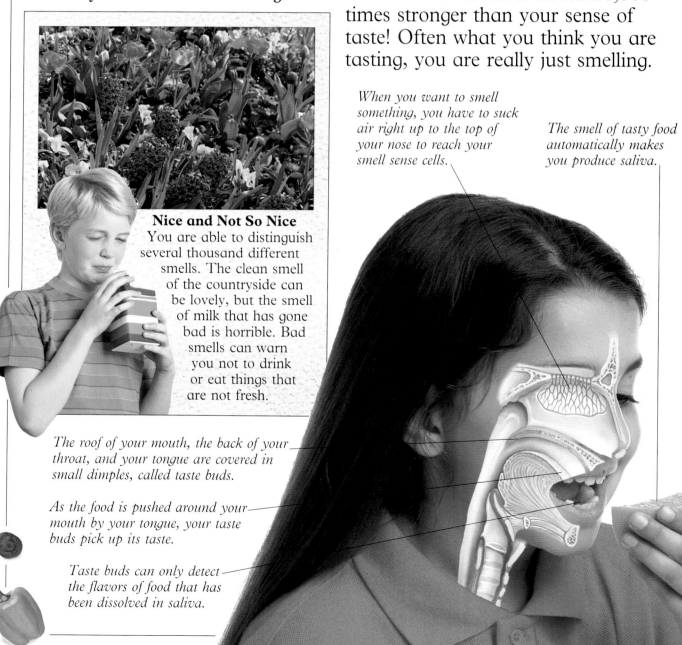

Nice and Not So Nice
You are able to distinguish several thousand different smells. The clean smell of the countryside can be lovely, but the smell of milk that has gone bad is horrible. Bad smells can warn you not to drink or eat things that are not fresh.

When you want to smell something, you have to suck air right up to the top of your nose to reach your smell sense cells.

The smell of tasty food automatically makes you produce saliva.

The roof of your mouth, the back of your throat, and your tongue are covered in small dimples, called taste buds.

As the food is pushed around your mouth by your tongue, your taste buds pick up its taste.

Taste buds can only detect the flavors of food that has been dissolved in saliva.

Trick Your Taste Buds

When you eat something, your sense of smell helps you get the flavor. Block your nose and taste a carrot and a cucumber. It is hard to tell the difference between them. If you have a piece of onion under your nose and eat mashed apple, you will think you are eating onion. This is why you cannot taste your food properly when you have a cold.

Coffee grains are bitter.

Lemon is sour.

Smoked fish is salty.

Honey is sweet.

Still Hungry?

The look of food is important as well as its taste and smell – pink corn still tastes like corn, but would you want to eat it?

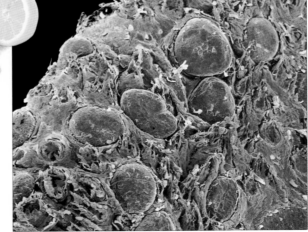

Busy Buds

If you look at a tongue through a strong magnifying glass, you can see lots of bumps. Around the bases of these bumps, there are taste buds. Inside them, there are special cells that sense taste.

Tongue Map

There are four sorts of taste buds. Each type is on a different part of your tongue and can detect a different taste. The four tastes are salty, sweet, bitter, and sour.

TEETH

You use your teeth to break up food into pieces that are small enough to swallow. Some are shaped for biting and others for chewing. You have two sets of teeth. The first set are called milk teeth, and there are 20 of them. At the age of about six, you start to lose your milk teeth. One by one, the second set, the 32 adult teeth, grow in their place. Teeth are strong and keep working for many years.

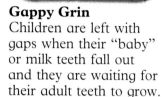

Gappy Grin
Children are left with gaps when their "baby" or milk teeth fall out and they are waiting for their adult teeth to grow.

Cases for Braces
Sometimes teeth grow crookedly in the mouth. Often this can be corrected by wearing teeth braces.

Healthy gums are just as important as healthy teeth – they help to hold your teeth in place.

Still Rooted
This skull shows how teeth are rooted into the two powerful jawbones.

Jawbone

Complete Set
A full set of adult teeth has 8 incisors, 4 canines, 8 premolars, and 12 molars. The 4 molars farthest back are called wisdom teeth.

Incisors

Canine   Premolars

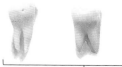

Molars

Inside Story

Your teeth are alive. The part that sticks out from your gum is the crown. The part under the gum is the root.

Enamel is the hardest substance your body makes. A thin layer of it covers all your teeth.

Nerve

Gum

Under the enamel, there is a layer of hard, bonelike material, called dentine.

Jawbone

The soft, middle part of the tooth is called the pulp. It contains nerves and blood vessels.

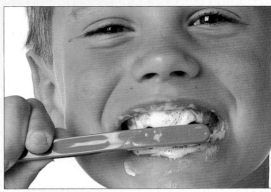

Brushing Up

Saliva and bits of food form a mixture, called plaque. If plaque builds up on your teeth, it can cause them to decay and even fall out. Brushing your teeth and gums after meals and before going to bed helps to remove the sugary foods that lead to tooth decay.

Your knobbly back teeth are called molars. They crush and grind food when you chew.

The sharp, chisel-shaped teeth at the front of your mouth are incisors. They can cut and bite tough food.

Sometimes people have extra teeth.

At about the age of 18, wisdom teeth may grow at the back of your mouth – but they don't make you wise!

There are molars on the bottom, too.

Premolars have two edges, or cusps. They tear and grind up your food.

Canine teeth are sharp and pointed. They are used for tearing food.

EATING

Food is the fuel that provides energy for your body. The energy is measured in units, called calories. Before your body can use the food you eat, it has to be broken down into tiny bits that are small enough to pass into your blood. This digestion takes about 24 hours as the food flows through a long tube winding all the way from your mouth to your bottom.

Squashed Insides
The small intestine is tightly coiled up inside your body.

1. Food starts being digested in your mouth. Your spit, or saliva, has a digestive juice that starts to break down the food.

Your liver is a 'chemical factory'. It also stores vitamins.

Windpipe

2. The food travels down a food pipe, called the esophagus.

3. Your stomach is a thick bag. Food is churned up inside it and mixed with strong stomach juices to make a kind of soup.

4. After leaving your stomach, your food flows down your small intestine. Nutrients from the food seep through the thin walls into your blood.

Large intestine

Small intestine

Fats

Minerals

Vitamins

Proteins

Get into Groups!
Foods can be put into groups. Fats and carbohydrates provide you with energy. Vitamins and minerals keep you healthy. Proteins build cells and so help your body to grow and repair itself.

5. Your large intestine holds the food that your body cannot digest. Later, it is passed out of your body through your rectum.

6. Your bladder stores urine. When it fills up, you feel the need to go to the bathroom to empty it.

A Lump in the Throat

You can swallow even if you are standing on your head! This is because your food does not slide down through you – it is squeezed along by muscles in your digestive tube. This is called peristalsis, and it happens all the time, without you having to think about it. The muscles of a snake can squeeze an egg through its body in the same way.

Walking uses about 240 calories an hour.

Sleeping uses about 65 calories an hour.

Drawing uses about 85 calories an hour.

Netball and other vigorous sports use about 550 calories an hour.

Fuel Burning

If a car travels very fast, it uses up more fuel than if it goes slowly. The same is true of your body. When you exercise, you use up more calories than when you are asleep.

You have two kidneys. Each one is about the size of your clenched fist.

Narrow tubes, called ureters, take urine from the kidneys to the bladder.

Carbohydrates

Any water your body does not need is turned into urine by your kidneys.

How Long?

If you could stretch your whole digestive system out in a straight line, it would be about as long as ten yardsticks!

MUSCLES

Try to sit as still as you can. Is anything moving?
Even when you think you are completely still,
many parts of your body are moving.
Your heart is beating, and your
intestines and lungs are
working. All these
movements are made by
muscles. You have over 600
muscles spread throughout your
body. Every bend, stretch, twist,
and turn you make depends on
them. You use about 200 muscles
each time you take a step, and
many more when you jump.

*Your brain sends
messages to your
muscles and makes
them move.*

*The largest muscle
in your body is the
gluteus maximus
muscle in your
thigh and bottom.*

*If you stand on
tiptoe, you can see
your calf muscles in
the back of your leg.*

Muscle Food
To keep your muscles working
properly, you need a diet that
includes protein. Foods that are
full of protein include eggs, cheese,
and dried beans.

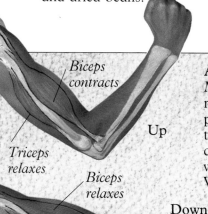

Biceps
contracts

Up

Triceps
relaxes

Biceps
relaxes

Down

Triceps contracts

Arm Bend
Muscles are attached to bones and
make them move. But they can only
pull. They cannot push – which is why
they always work in pairs. In your
arm, the biceps and triceps muscles
work together to move it up and down.
When the biceps pulls, or contracts,
it gets shorter and fatter
and bends the arm.
As the biceps pulls, the
triceps muscle relaxes.

*Before you begin to
make any strenuous
movements, you should
always warm up your
muscles by doing gentle
loosening-up and
stretching exercises.*

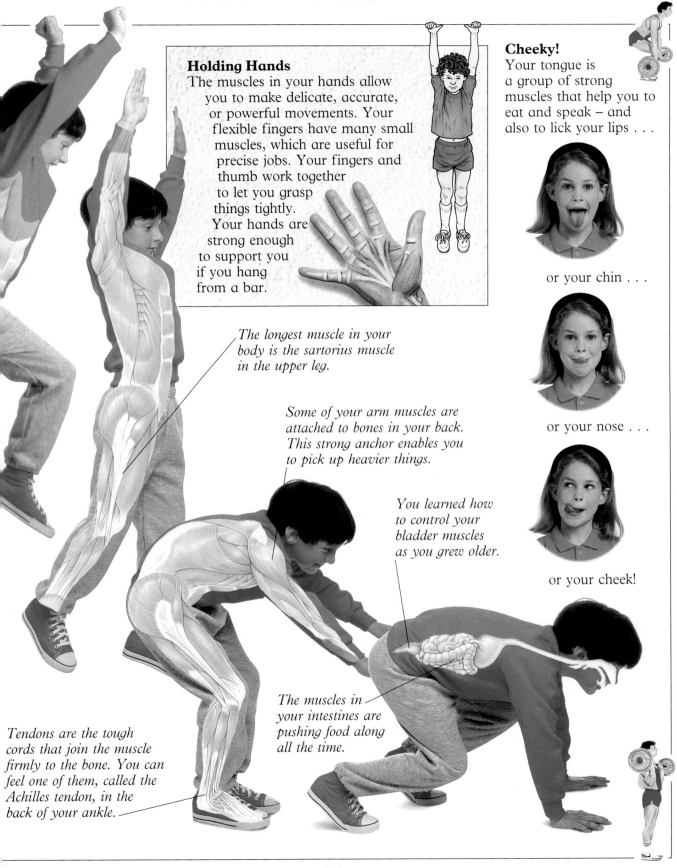

Holding Hands
The muscles in your hands allow you to make delicate, accurate, or powerful movements. Your flexible fingers have many small muscles, which are useful for precise jobs. Your fingers and thumb work together to let you grasp things tightly. Your hands are strong enough to support you if you hang from a bar.

Cheeky!
Your tongue is a group of strong muscles that help you to eat and speak – and also to lick your lips . . .

or your chin . . .

or your nose . . .

or your cheek!

The longest muscle in your body is the sartorius muscle in the upper leg.

Some of your arm muscles are attached to bones in your back. This strong anchor enables you to pick up heavier things.

You learned how to control your bladder muscles as you grew older.

Tendons are the tough cords that join the muscle firmly to the bone. You can feel one of them, called the Achilles tendon, in the back of your ankle.

The muscles in your intestines are pushing food along all the time.

SKELETON

Without a frame to support your body, you would collapse, lose your shape, and be unable to move. Your body's frame is called a skeleton. It gives your body strength, and it protects the soft parts inside. Your skeleton is made up of more than 200 bones. They are light enough to allow you to move about easily, and they have joints so that you can bend your body to do many things.

You have 12 pairs of ribs. They are all joined to a row of bones in your back, called your spine.

From the side, your spine looks curved, like the letter S. It helps you to stand up straight.

Ulna

Femur

A Tall Order
Your bones keep growing until you are in your early twenties. You cannot change your height – it is decided in your genes and passed on from your parents. But you are about half an inch shorter in the evening than you are in the morning! This is because the pads of cartilage in your spine get squashed as you walk about all day.

Yes and No Bones
The bones of your spine are called vertebrae. The top two vertebrae, the atlas and axis fit together to allow your head to nod and to move from side to side.

Atlas

Axis

Your nose is no made of bone but o, rubbery material, called cartilage. If you look a a skeleton, you will not see a nose bone, only nose holes.

Radius

Fibula

Tibia

Your ankle is a joint. It is made up of bones in the foot and the ends of the leg bones, the tibia and the fibula.

Soft Center

Some animals, like this crab, do not have a skeleton inside them. Instead, they have a hard outer covering, called an exoskeleton.

Inside Information

Your bones are all hidden inside your body. So if doctors want to look at them, they have to take special photographs, called X rays. The X-ray camera can see straight through your skin and show what the bones look like. On this X ray of a hand, you can see that the bone connected to the little finger is broken.

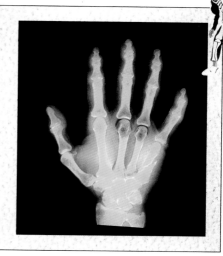

Bones give muscles a place to hang on to, but without these muscles, the bones would not be able to move. Muscle power is transferred to the bones along strong bands, called tendons.

Your hip joint is where the end of the thighbone, or femur, fits into a socket in your pelvis. This joint helps you to bend your body almost in half.

Your basin-shaped pelvis supports the upper half of your body and also protects soft parts, such as your bladder.

Your spine can only curve gently. If it bent any farther, it would damage your spinal cord – the nerve cable that carries messages to and from your brain.

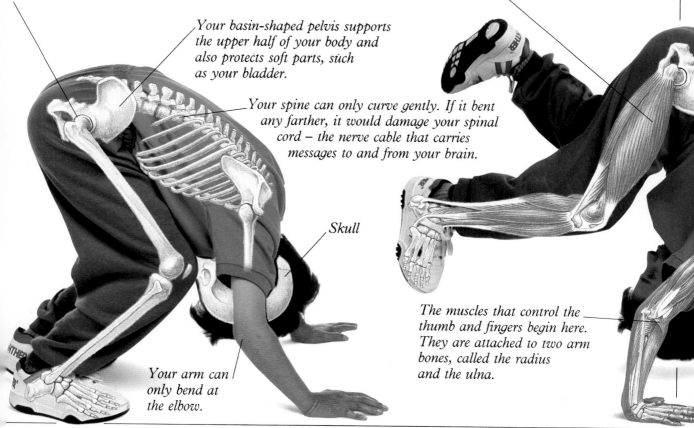

Skull

Your arm can only bend at the elbow.

The muscles that control the thumb and fingers begin here. They are attached to two arm bones, called the radius and the ulna.

BONES

Your bones are hard and strong. They are not solid, though, so they are not as heavy as you would think. In fact, they only make up 14 percent of your total body weight – they are lighter than your muscles. Bones are not dead and dry – they are living, and can repair themselves if they break. Your body is made up of lots of bones all working together and linked by joints. If you had no knee joints, you would have to walk with stiff legs.

What's inside a Bone?

The outer part of all your bones is hard and tough, but the inside of many of them is spongy. These lightweight, soft centers are crisscrossed by small struts, which make your bones strong but not too heavy. This idea of strength without weight is copied in buildings such as the Eiffel Tower.

Crisscross struts

Your two feet contain one-quarter of the bones in your whole body!

The spongy inner bone looks like a honeycomb.

Some bones are filled with jellylike marrow. Red blood cells are made in the bone marrow.

Blood vessels take oxygen and food to bone cells.

Your neck is much shorter than a giraffe's, but it has the same number of vertebrae!

The tiny tail bones at the end of your spine, called the coccyx, protect your spinal cord.

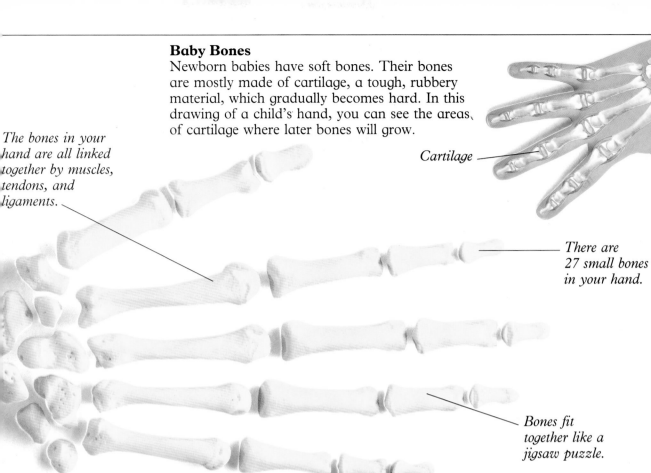

Baby Bones
Newborn babies have soft bones. Their bones are mostly made of cartilage, a tough, rubbery material, which gradually becomes hard. In this drawing of a child's hand, you can see the areas of cartilage where later bones will grow.

Cartilage

The bones in your hand are all linked together by muscles, tendons, and ligaments.

There are 27 small bones in your hand.

Bones fit together like a jigsaw puzzle.

Bone Work
You have three kinds of bones – long bones, like those in your legs; short bones, such as those in your hand and spine; and flat bones, like your shoulder and skull. Bones are linked by different kinds of joints, which allow them to move in different ways.

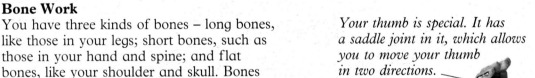

Your thumb is special. It has a saddle joint in it, which allows you to move your thumb in two directions.

You have flat gliding joints in your foot.

Your kneecap, or patella, protects your knee.

The shoulder has a ball-and-socket joint. The round end of one bone fits into a cup-shaped hole in the other. Your shoulder can move in a complete circle.

Your knee joint, like your elbow, is a hinge joint. The end of one bone fits into a sort of hollow in the other. This kind of joint will only bend in one direction.

WHERE DO I COME FROM?

You began your life as an egg that was only about the size of this period. This tiny fertilized egg grew for about nine months inside your mother before you were born. While a baby is growing, it relies on the mother for everything, and although a baby cannot do very much when it is first born, already it is a complete and very special person.

The Beginning

Everyone is made of billions of living units, called cells. A baby starts when a single egg cell from a woman and a single sperm cell from a man join together to make one new cell, called a fertilized egg. The egg attaches to the wall of the woman's uterus and begins to grow by dividing to form new cells. Even though the uterus looks small in this picture, it is very stretchy, like a balloon. It will expand as the baby grows bigger.

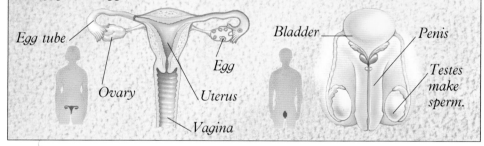

Egg tube

Ovary

Egg

Uterus

Vagina

Bladder

Penis

Testes make sperm.

The fertilized egg grows and divides in two.

It divides into 4, then 8, then 16, and so on . . .

Sperm are like tiny tadpoles with long tails. They swim to the tube where the egg is waiting. One sperm may then fertilize the egg.

While it is dividing, it is traveling to the womb, or uterus.

After about eight weeks, the group of cells starts to look more like a baby.

The fingers, toes, and face are formed.

After nine days, the egg attaches itself to the wall of the uterus.

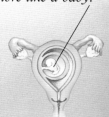

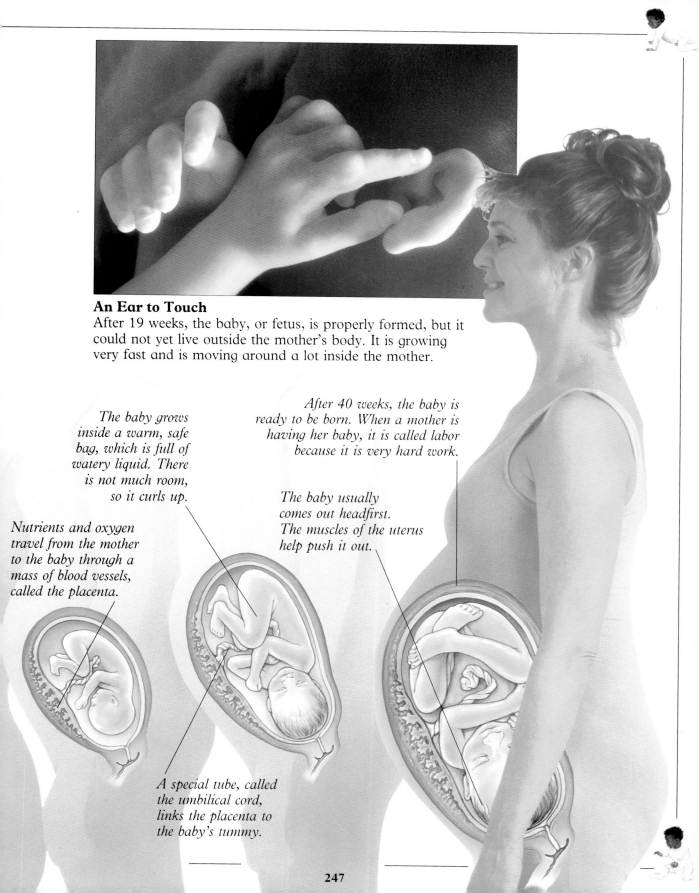

An Ear to Touch

After 19 weeks, the baby, or fetus, is properly formed, but it could not yet live outside the mother's body. It is growing very fast and is moving around a lot inside the mother.

The baby grows inside a warm, safe bag, which is full of watery liquid. There is not much room, so it curls up.

After 40 weeks, the baby is ready to be born. When a mother is having her baby, it is called labor because it is very hard work.

Nutrients and oxygen travel from the mother to the baby through a mass of blood vessels, called the placenta.

The baby usually comes out headfirst. The muscles of the uterus help push it out.

A special tube, called the umbilical cord, links the placenta to the baby's tummy.

CHAPTER 3

OUR WORLD

Early people hunted for food only when they were hungry. Later, they learned to farm, growing crops and keeping animals so they had a more regular supply of food. Today, most of the food in stores and supermarkets still comes from farms – meat and dairy products, grains and rice, fruits and vegetables. However, machines now do most of the work, and special chemicals kill pests and help crops grow.

As their lives became more settled, our ancestors formed into communities and had more time to enjoy themselves. The simple entertainments of singing, dancing, and drawing branched out over time into plays and, much later, films and television.

Today, whether they live on high mountains, dry deserts, or swampy marshlands, all men and women have to feed, clothe, and house themselves. But from country to country, people do these things in an enormous variety of ways.

People in the Past
Arts and Entertainment
Food and Farming
People and Places

PEOPLE IN THE PAST

The lives of men and women who lived a long time ago were very different from ours. The first people made their homes in caves and had to kill their food or find it growing wild. For a long time, no one could read or write, so information and stories had to be learned by heart in order to be passed on.

Changes in the way people lived came gradually, when travelers – traders, or soldiers who went to war in foreign lands – took new ways of doing things, new foods, or new materials from one place to another. Sometimes, changes were particularly important, like the invention of the wheel or the printing press. These changes tended to spread quickly and alter everyday life completely.

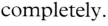

Viking warrior

18th-century gold doubloons

Ancient Peruvian pot

Pottery beaker ca. 2200 BC

19th-century
Russian
cossack
pistol

17th-century samurai sword

Egyptian wall painting ca. 1400 BC

Illuminated manuscript

Arapaho chief's
headdress

Ancient
Egyptian
reed pens

Beaded
Blackfoot
moccasins

HUNTERS AND GATHERERS

Horse

The earliest people lived by hunting and gathering their food – meat, fish, vegetables, and fruit. They moved with the seasons, sheltering in caves or tents. The animals they hunted gave them food to eat, fat to fuel lamps, skins to make tents and clothes, and bones to make weapons, tent supports, jewelry, and playthings.

Some European hunter-gatherers, who lived 16,000 years ago, painted animal pictures in caves. We don't know exactly why they did this, but cave painting may have been part of their religion or a kind of magic.

Making Flint Tools

Flint is a stone that is easy to work and can be given a sharp edge. The hunter-gatherers used it to make tools and weapons. A pebble or bone hammer was used to strike long flakes of flint from the main stone. The flakes were shaped and then the edges were chipped to make them razor sharp.

The mammoth was like a modern-day elephant, but covered in fur. It was hunted for its meat, skin, and tusks.

An ax with a flint blade

A spear with a bone tip

Animal Magic

Beautiful pictures of the animals hunted by these people were painted on the walls and ceilings of caves in southwest France and in parts of northern Spain.

Bison

Reindeer

The hunters hurled heavy stones at their prey.

The spears used for hunting were often made from a flint or bone arrowhead tied onto a wooden shaft.

Unusual Decoration

Teeth and bones from animals were made into pieces of jewelry, such as this necklace.

The hunters wore clothes made from animal skins.

First Fashion

Animal skins were used to make clothes. First the skins were stretched out, held in place with pegs, and then scraped to clean them and make them soft. Next, the skins were cut to shape. The pieces were sewn together using needles made from animal bone and long, thin strips of hide.

A thong made from animal hide

Skin scrapers

A knife used to cut animal hide

THE EARLY FARMERS

Bullock

Goat

Sheep

About 12,000 years ago, there was an important change in the way people lived. They began to settle in one place and to farm the land. This happened in the Zagros Mountains in present-day Iran. Here the people discovered how to grow crops for food, and how to tame wild animals to give them meat and milk and to carry heavy loads.

Because the people no longer moved about, they built houses to live in. They learned to spin and weave cloth and also to make pottery.

Taming Animals
Young wild animals were caught by hunters and raised on the farms so they became used to being around people.

Bringing in the Harvest
When the crop was ripe, it was cut down with a sickle. This was made of a sharp flint blade set in a wooden handle.

First Crops
This is emmer wheat. Emmer was developed from the seeds of wild grasses to become an important food source.

Before it is ground, this ear of corn will be beaten to separate the grains from the husk.

Coiling Clay

This girl is making a pot by coiling long, thin rolls of clay on top of a flat clay base. She is making the pot in the same way the first potters did, 9,000 years ago.

The grain was ground between two heavy stones and made into flour.

The flour would be mixed with water and made into round, flat loaves.

From Farm to Town

Farms attracted people. As they grew larger, villages and towns were formed. The first known town, Çatal Hüyük, was built in Turkey about 9,000 years ago.

The walls of the houses were made of mud. Because of this, the houses were not very strong and could only have a few windows.

The roofs were made of branches, reeds, and straw covered in mud.

The houses touched each other, and there were no streets.

People got into their houses by climbing a ladder and going through a hole in the roof.

THE SUMERIANS

About 7,000 years ago, farmers began to move into an area of land between the Tigris and Euphrates rivers. This fertile land was called Mesopotamia, in what is now Iraq. In the south of Mesopotamia was the land of Sumer. The Sumerians were very inventive. They developed the first form of writing and recording numbers, and invented the wheel and the plow.

The Sumerians grew bumper crops of cereals, which they traded for things they needed: wood, building stone, or metals. Wheeled carts and their skills in writing and using numbers helped them develop long-distance trade.

A merchant makes a record of the goods he has sold.

How Writing Began
The Sumerians drew pictures on soft clay with a pointed reed. The pictures were drawn downward in lines, from the right-hand side.

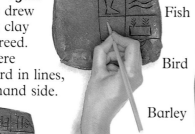

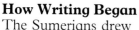

Later, they started to write across the tablet, from left to right. The reed tip became wedge shaped, as did the marks it made.

Fish

Bird

Barley

Ox

The signs were then joined together to build up words and sentences.

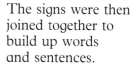

In time, the picture signs changed so much that the original objects were hard to recognize. This writing is called cuneiform, which means "wedge shaped."

It's a Deal
Instead of writing their names, the Sumerian traders used a seal to sign their contracts.

Flooded Fields
Mesopotamia was watered by the Tigris and the Euphrates, which flooded in the spring. The farmers dug out basins and canals so the river water could be stored and used to water fields away from the river. Because of this, the land was fertile and harvests were good.

Wood was imported from Syria and Lebanon.

Wheeled carts were able to carry heavy objects over long distances.

Pots were used to store grain and oil.

Crosspiece

Spokes

Tilling the Soil
The first plows were made of wood. Later, the blade was made of bronze. Despite the flooding from the rivers, Sumer was baked hard in hot weather. The plow made it possible to break up the hard-to-work soil.

Revolutionary Invention
The first wheels were made of planks of solid wood held together with crosspieces.

They were clumsy and heavy. In time, lighter wheels were made. These had many spokes.

THE EGYPTIANS

The Egyptians believed that when they died, they went on to another, everlasting life. To live happily in the afterlife, they needed their earthly bodies, and great care was taken to ensure that they arrived in style. The dead bodies were preserved in a special way, called mummification.

To mummify a body took a long time, as much as 70 days. First, parts of the body were taken out and put in tightly sealed jars. Next, the body was dried out by covering it with natron, a white powder like salt. It was left for 40 days, rubbed with sweet-smelling oils, and then covered in molten resin. Finally, the body was wrapped in linen to make a neat package.

Well Kept
This is how the mummy of King Seti I looks underneath all his bandages. He was buried at Thebes over 3,000 years ago. Despite his age, you can still see his face very clearly.

A picture of the dead person is painted on the mummy case or coffin.

On this coffin, the arms are shown crossed over the chest.

The coffin is made to follow the shape of the body inside.

These little drawings are called hieroglyphs. Each picture means a word or sign in the ancient Egyptian language.

Royal Tombs
Egyptian kings, called pharaohs, were buried in tombs known as pyramids.

Creatures Great and Small

The Egyptians believed in many different gods and goddesses, some of which took the form of animals. They mummified these creatures, as well as people. When treated, they made some odd shapes.

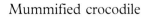

Mummified crocodile

Mummified cow

Mummified cat

Red straps painted on the coffin show that the mummy was a priestess.

The lid of the coffin is decorated with symbols of the gods. These are the wings of the sky goddess, Nut.

This coffin is made of wood. Early ones were made of clay or woven out of reeds like a basket.

The hieroglyphs may have been spells to help the priestess on her journey to the next life.

Brightly colored figures and symbols are painted on the inside of the coffin.

This mummy case may have been one of a nest of coffins, each fitting inside the next like a set of Russian dolls.

All Wrapped Up

All mummies were wrapped up tightly in lots of material. As much as 450 square yards (375 square meters) of linen might be needed to wrap up a single mummy.

GREEK GAMES

All over ancient Greece, festivals were held in honor of the Greek gods. They included competitions in sports, music, and drama. The most famous of the festivals was the Olympic Games, an event first held in 776 BC. There were no team races, and the male athletes competed as individuals. Their prize was a simple wreath of olive leaves, but if you won, you became a hero.

The discus was made of stone or bronze.

The athletes competed barefoot and wore no clothes.

Liftoff!
The long jump was the only jumping event included in Greek athletics.

Fighting Fit
Athletics training kept men fit for war. The connection between sports and war is shown in the race-in-armor event.

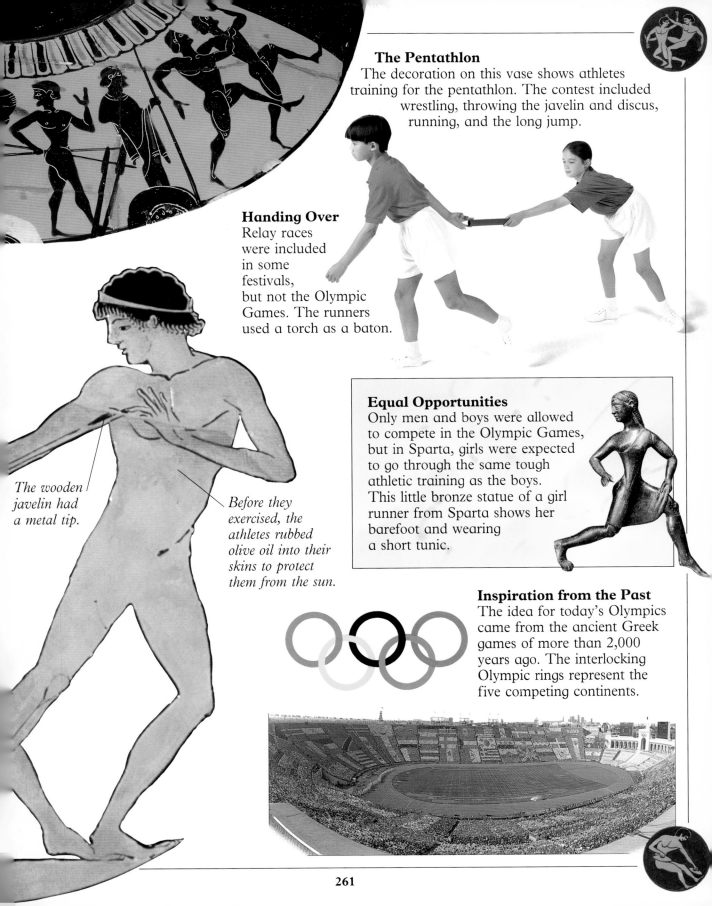

The Pentathlon

The decoration on this vase shows athletes training for the pentathlon. The contest included wrestling, throwing the javelin and discus, running, and the long jump.

Handing Over

Relay races were included in some festivals, but not the Olympic Games. The runners used a torch as a baton.

The wooden javelin had a metal tip.

Before they exercised, the athletes rubbed olive oil into their skins to protect them from the sun.

Equal Opportunities

Only men and boys were allowed to compete in the Olympic Games, but in Sparta, girls were expected to go through the same tough athletic training as the boys. This little bronze statue of a girl runner from Sparta shows her barefoot and wearing a short tunic.

Inspiration from the Past

The idea for today's Olympics came from the ancient Greek games of more than 2,000 years ago. The interlocking Olympic rings represent the five competing continents.

ROMAN LIFE

The Romans planned their towns to include magnificent public buildings such as temples, the town hall, baths, and places of entertainment. There were also grand houses for wealthy families. But side by side with these were the tumbledown dwellings where most people lived – overcrowded apartments built over shops and workshops. They had no toilets or kitchens, and the poor cooked outdoors or bought hot food from stalls. Since the Romans lived, traded, and ate in the streets, their towns were noisy places. Traffic was bad, too, with carts and wagons bringing country goods through the busy streets.

In the Kitchen
This scene shows what a typical Roman kitchen was like. You would have found it in the town house or villa of a rich family.

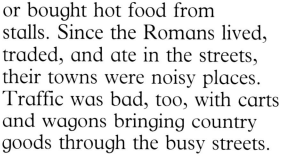

Country Living
Fine houses, called villas, were built on estates in the countryside.

Building being built of bricks

Wharf

Fishing nets drying

A wooden bridge over the river

Herbs dried over the stove. The Romans liked their food highly flavored. Herbs and spices also helped make food keep longer.

Bread was an important part of Roman diet and the basic food of the poor.

A saucepan made of bronze

Storage jars, called amphorae, were used to hold wine.

Slaves did the housework for wealthy Romans.

Jugs for serving wine

A sailing boat carries cargo from the port.

Cats were kept as pets for children and to chase away mice and rats!

The floor was centrally heated. This worked by sending warmed air through pipes laid under the floor.

Modern Conveniences
Unlike the poor in their cramped dwellings, wealthy Romans lived in well-planned houses with lots of home comforts. There was running water, a toilet and bathroom, a kitchen, and even central heating. In many ways, a Roman house was not unlike the one you live in today.

RAIDERS FROM THE SEA

Late in the eighth century, a seafaring people from the countries now known as Denmark, Norway, and Sweden began to sail abroad in search of new homelands. They were called Vikings. In their amazingly fast and adaptable longships, Vikings were the best sailors in Europe. They often behaved like pirates. Then, gradually, the Vikings began to settle in the lands they had once raided, although they continued to be great sea traders. Many of the Vikings became Christians. It was the Vikings who founded the cities of Dublin in Ireland and Kiev in the Ukraine. The Normans were descended from Viking settlers in northern France.

The mast supported a big square sail. This was used when the longship was out at sea and there was plenty of wind to fill the sail.

Longships were called "serpents of the sea." They were fitted with fearsome animal carvings, called prowheads.

As Vikings prepare to land on a foreign shore, warriors stand at the front of the longship, with settlers in the middle.

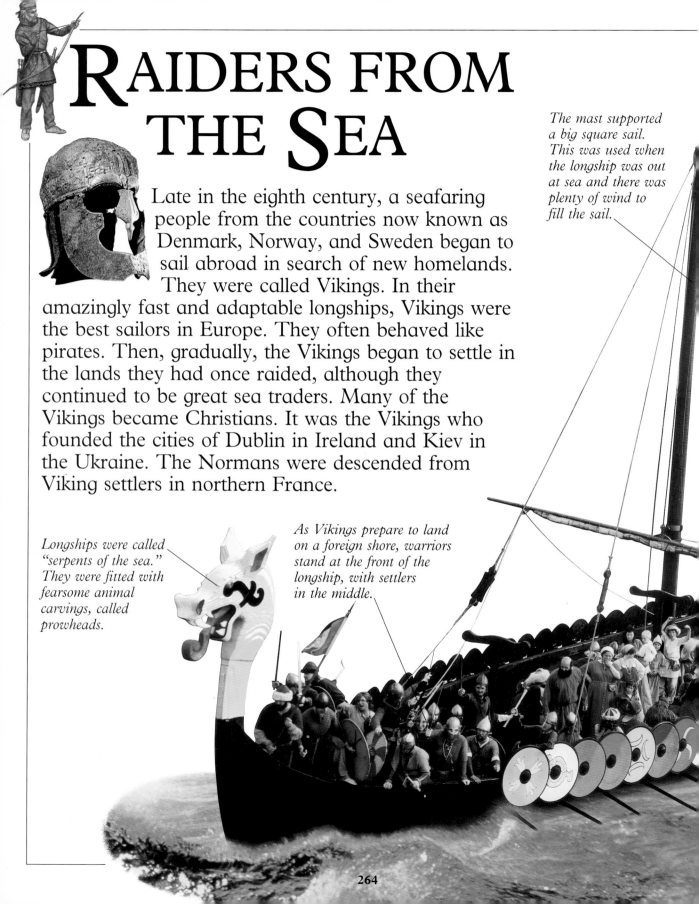

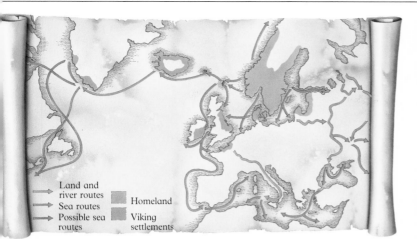

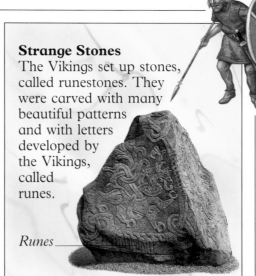

Strange Stones
The Vikings set up stones, called runestones. They were carved with many beautiful patterns and with letters developed by the Vikings, called runes.

Runes

Where the Vikings Went
The Vikings roamed great distances. Merchants traveled to Russia, Iran, and around the Mediterranean Sea. They traded such northern things as furs and walrus tusks for southern things, such as silk and silver. Explorers sailed to Europe, Iceland, Greenland, and also Newfoundland in North America. They called this Vinland.

Map key:
→ Land and river routes
→ Sea routes
→ Possible sea routes
☐ Homeland
☐ Viking settlements

Longships were made of overlapping planks of wood. They were very light and could be carried overland for short distances to get from one river to another.

Fascinating Tales
Storytelling was very popular. These stories, or sagas, were long poems telling of brave deeds, journeys to strange lands, and victories in battle. When people couldn't read or write, they remembered stories by learning them by heart.

THE CRUSADES

The city of Jerusalem and the Holy Land of Palestine (now Israel and Jordan) are special to people of three faiths – Jews, Muslims, and Christians. About 900 years ago, several wars were fought in the Middle East for control of Jerusalem. Muslims seized the city and refused to allow in Christians. As a result, European Christians – called crusaders from the Spanish word *cruzada*, meaning "marked with a cross" – set out to take Jerusalem.

After four years of struggle, the crusaders captured Jerusalem in 1099 and held it for nearly 100 years. Then it was recaptured by a great Muslim leader called Saladin. More crusades followed, but none was successful. However, the returning crusaders brought many new objects and ideas back to Europe.

This crusader knight is wearing armor called chain mail.

Over his armor, the knight is wearing a loose white coat with a cross on it – the sign of a crusader.

Stronghold
The crusaders built strong castles in the Arab style to defend the land they had captured. This is Krak des Chevaliers in Syria.

The leg pieces are made of chain mail.

Body Beautiful

The crusaders took back to Europe new customs that they learned in the Middle East. Cosmetics such as rouge to redden the cheeks and henna to color hair became common for women. Glass mirrors replaced polished metal discs. Perfumes to scent clothes and the body were used. Being clean became popular!

Roman numerals

I II III IV V VI VII VIII IX

Arabic numerals

0 1 2 3 4 5 6 7 8 9

One, Two, Three . . .

The Muslims developed a series of numbers, called Arabic numerals, that were much easier to use than those of the old Roman system. They are used today throughout the world.

Bookworms

Arab learning was more advanced than that of the crusaders. They had lots of books and libraries at a time when there were very few in Europe. The crusaders took many books back with them.

This puppet of a Muslim Arab is fighting a Christian knight. For centuries, puppets have been used to act out folktales, religious stories, and famous battles and scenes from history.

Healing Arts

Muslim doctors were very good surgeons and were skilled in using plants and herbs to make medicines. They used opium and myrrh to ease pain when people were having operations.

Musical Instruments

The Muslims invented a musical instrument called the al'ud, which was called the lute in Europe. The modern guitar was developed from this instrument.

A Muslim foot soldier usually carried a round shield for protection. It may have been made of wood or of layers of hardened leather sewn together.

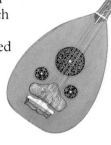

Great Inventors

The Muslims made fine scientific instruments like this astrolabe. It used the positions of the stars and planets to show travelers which way to go in the empty desert.

Boots made of leather or felt were the most common footwear for a Muslim soldier.

MARCO POLO

Making a Swap
The Venetian merchants exchanged jewels, silver, and gold for goods from the East that were highly valued in Europe. These included spices, silks, porcelain or "China," and fine carpets.

In 1271, a merchant called Marco Polo and his family set off from Venice on an extraordinary journey to China. They traveled along the Silk Road, an important trading route between Europe and the Far East. Merchants had been using the route for over 1,700 years before the Polos, but they were the first to travel its whole length. In an age when there were no planes, trains, buses, or cars, they crossed thousands of miles of mountains, deserts, and plains on foot, on horses, and even on camels. Their journey to China took them over three years. The Silk Road could be very dangerous, and the traders were sometimes attacked by bandits. Because of this threat, they traveled together in groups.

Wealthy Traders
Venice was the most important port in Europe. Its merchants traveled by sea and land to bring back goods from the East to sell to other towns in Europe.

Camels were used by merchants when crossing the desert regions of central Asia. Unlike horses, they were able to travel long distances without needing water.

Besides the barren deserts, the merchants traveling on the Silk Road had to climb over fearsome mountain ranges and cross flooded rivers.

Pepper

Cloves

Cinnamon

Mace

Nutmeg

In the desert, travelers had to take all their own food and water with them.

A camel train was part of a caravan, a large group of people who traveled together.

Silk tunic

Persian carpet

Porcelain jar

The camel driver is covered from head to foot in heavy clothing to protect him from the harsh winds of the desert.

In the Court of Kublai Khan

When the Polos arrived in China, they found it under the control of the Mongol emperor Kublai Khan. Marco Polo stayed at the Mongol court for 17 years and became a trusted advisor to Kublai Khan.

THE RENAISSANCE

In the 15th century, Europe was bursting with new ideas about art, learning, and religion. Many of the people who could read and write began to ask questions and to do experiments for themselves rather than follow what their rulers and priests told them to think. They rediscovered many of the ideas that the Greeks and Romans had about life and the world. As a result, this period became known as the "Renaissance," a French word meaning "rebirth." The Renaissance started in the cities of northern and central Italy but gradually spread all over Europe.

Before the Renaissance, most paintings showed mythical scenes or Bible characters, and everything in them looked perfect.

For the first time, Renaissance artists used live models and made them look like ordinary human beings.

A letter of type used in the printing press

The Printed Word
Before the printing press was invented by Johannes Gutenberg, books had to be copied by hand. Only the rich could afford them.

Everyday objects were used as props in the paintings.

The apprentice is grinding colors to make paints for his master. He will mix the powder with oil to produce oil paints.

The Gutenberg Bible
These pages come from one of the three books that made up the Gutenberg Bible. It was printed in the 1450s and was much admired for its quality.

In the 15th century, artists began to paint on canvas using oil paints. Before then, they had mainly used water-based paints.

Artists were no longer unknown craftsmen, and their names became famous throughout Europe during the Renaissance.

The Quest for Knowledge

During the Renaissance, scholars and scientists began doing experiments and inventing things to find out more about the world they lived in.

With the invention of the telescope, faraway planets could be seen in detail for the first time.

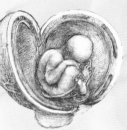

The artist Leonardo da Vinci was interested in the workings of the human body. This is one of his drawings.

This model was made from a design by Leonardo, showing an early kind of tank.

Leonardo's creative imagination led him to produce designs for a flying machine.

FOOD FROM THE NEW WORLD

From 1492 onward, European explorers sailed across the Atlantic to what they called the New World of North, Central and South America. They discovered a treasure trove of gold and silver, but they found other treasures, too. These were foods that grew only in the New World, such as corn (maize) and potatoes, and plants that could be made into medicines. In fact, you may be surprised to find out how many of the foods you eat originated in the Americas. Every time you eat a tomato or some mashed potato or have chocolate for a treat, just think: these foods have been enjoyed in the Americas for hundreds of years.

Sunflower

Cocoa

A World of Golden Treasure

The Indian peoples of America had vast quantities of gold, which they used to make into jewelry. The Europeans plundered most of this treasure.

Maize was eaten boiled, roasted, or ground into flour. The explorer Christopher Columbus took it back to Europe when he returned from his voyages.

This is corn meal, made out of ground corn. It was used to make breadlike foods.

Peanut

Native American Plants

These include cocoa (for chocolate), sunflowers (for cooking oil), and peanuts.

Healing Plants

When the explorers reached the Americas, they found skilled healers among the people living there. These healers used thousands of plants to make medicines. Their remedies were used to cure many illnesses, including stomach pains, headaches, coughs, and fevers. Many of these plants are still used today to make medicines.

Quinine is used to prevent an illness called malaria. It is made from the bark of the cinchona tree.

Cinchona leaf

Cinchona bark

Quinine tablets

"Pyne fruit," or pineapple, was one of the new fruits found by the explorers. Pineapples were also used to make wine.

Black beans

Lots of different beans came from Mexico. These are red beans.

These small, hot peppers are chilies. They were used to flavor the bowls of maize porridge that the Mexicans ate for breakfast.

Sweet potatoes

Tomatoes started off as weeds growing in the maize fields. In time, different kinds were grown to eat. The first ones to reach Europe were probably yellow.

Avocados look like pears with rough, tough skins. They were first grown in Central America.

Potatoes were first grown in the Andes Mountains. They were loaded aboard the treasure ships as food for the sailors.

THE LIFE OF THE SAMURAI

For about 700 years, the Samurai were the honored knights of Japan. Fierce fighters, they had a tough training, becoming experts in fencing, wrestling, archery, and acrobatics, and they had a special code of behavior. The word "samurai" means "one who serves," and any Samurai worthy of the name was absolutely loyal to his lord, ready to obey any command without question.

But although the Samurai were professional fighters, away from battle they were not violent men. The Samurai believed in Zen Buddhism, a religion that taught respect for all living things. The Samurai were also taught to love art and learning, taking pride in their skill at painting, writing poetry, and even flower arranging.

The large horned helmet was meant to terrify enemies as much as to protect its wearer.

Western Merchants
When the Portuguese arrived in Japan in 1543, they brought guns with them. The guns changed the nature of warfare in the country completely.

Women and Ladies

Just as the Samurai obeyed his lord, so the women of his own family had to obey him. Graceful, musical, and artistic, these Samurai ladies were expected to make homes for their lords and masters.

By contrast, the peasant women who labored in the fields had to work as hard as beasts of burden.

A Samurai's most important weapons were his two razor-sharp swords, one long, the other short. At birth, every Samurai boy was presented with a sword, which he kept for life.

A Samurai's armor consisted of six main pieces: the helmet, the face mask, the breastplate, the sleeves, the shin guards, and the loin guard.

Knowing Their Place

Life in Samurai Japan was strictly organized. From birth, everybody had a fixed place in society. Samurai families belonged to the upper classes.

The godlike Emperor was the official ruler, but the Shogun, his chief general, was really the most powerful person in Japan.

Shogun

Daimyo

The Daimyo were the nobles of Japan, and they were supported by Samurai warriors. They preferred to have nothing to do with money or the buying and selling of goods.

Samurai

Merchants and traders were not given much respect, in spite of their wealth.

Merchants

Lowest of the low were the peasants. They worked on the farms of the Daimyo and were treated like slaves.

Peasant

NEWCOMERS TO A NEW LAND

The Europeans who began arriving in North America in the beginning of the 17th century were traders and settlers rather than soldier-conquerors. At first, the contact between them and the people already living there, whom they called Indians, was friendly. The Native Americans showed the newcomers how to hunt, fish, and farm in a land of plenty. In return for their help and animal furs, the Native Americans were given objects such as knives, needles, fishhooks, and cloth. But before long, the settlers were taking more and more land for themselves and trying to change the ways of the Indians.

Warm furs from the forest animals of North America were taken back to Europe and sold for high prices.

European weapons, tools, and machines completely changed hunting and warfare for the Indians.

European cloth was prized for its bright colors and silkiness.

Creek
(Southeast)

Iroquois
(Northeast)

Tlingit
(Pacific Northwest)

Hidatsa
(Plains)

Hopi
(Southwest)

Sauk
(Great Lakes)

Paiute
(Great Basin)

The settlers gave the Indians silk thread and glass beads. The Indians used feathers and animal teeth to decorate themselves.

Warriors and Hunters

The work that men and women did varied from tribe to tribe, but usually the men were the hunters and warriors, while the women were the farmers and homemakers. Most Indians wore their hair long, and they enjoyed decorating their bodies and clothes.

Tepees, Longhouses, and Pueblos

There was great variety in the lives of the Native American peoples. How they lived – their clothes, their food, their religious beliefs – depended on the land and the weather. Some of the differences between the tribes are shown by their homes.

For comfort and protection against ants, snakes, and other dangers, the Indians wore slipper-like leather shoes, called moccasins.

Sweet corn, squash, and pumpkins were among the exotic American foods that the Europeans tasted for the first time.

Tlingit cedar-plank house

Plains Indian animal-skin tepee

Sauk mat-covered dome lodge

Paiute brush and reed encampment

Iroquois wooden longhouse

Hopi stone and sun-baked mud-brick pueblo

Creek storehouse

THE ASHANTI KINGDOM

Ghana

The Ashanti kingdom flourished for 200 years after its emergence in the 17th century in what is now Ghana in West Africa. The Ashanti were a highly organized people: the king had his own civil service, which carried out his commands throughout the country. The Ashanti were also fine warriors, and much of their wealth was based on selling slaves from the prisoners they captured in battle. They had vast quantities of gold, which was used to make jewelry and to decorate musical instruments and weapons. Ashanti goldsmiths were highly skilled in their craft. They used a special method to cast the metal, called the lost-wax technique.

The item to be cast in gold was modeled in melted beeswax.

The wax was made into thin sheets. These were cut into strips, which were used to make the model.

Clay was molded around the model, and a hole was made in the clay. As the mold was baked, the wax melted and poured out through the hole. Molten gold was then poured into the space.

When the metal had cooled and hardened, the mold was smashed open. The gold object was taken out and cleaned up.

Beautifully Made
These royal sandals have flowers of gold sewn onto them.

Ashanti goldsmiths worked gold into all kinds of objects. The handle of this sword was covered in a fine sheet of gold.

Hub of the Kingdom

Kumasi, the capital, was a teeming, bustling city. Many sumptuous parades and celebrations were held there.

Kente cloth

Fine Cloth

The Ashanti were experts at making beautiful cloth. One, known as kente, was made from cotton woven into narrow strips that were then sewn together. Adrinka was another sort of cloth. Large pieces of material were printed using stamps dipped in dyes. Patterns were built up in blocks or panels.

Printing Blocks

The stamps used to print cloth were made out of the shell-like fruits of calabash trees.

Elephant symbols were used in Ashanti jewelry to show the power of the wearer.

Dripping with Gold

The Ashanti wore rings on their fingers as well as gold bracelets on their wrists and at their knees.

CATHERINE THE GREAT

Catherine II, Empress of Russia during the 18th century, was called "Great" because she worked hard as ruler of her vast empire. She asked the advice of many of the major thinkers in Europe, and because she was interested in education, she started the Russian school system.

Catherine also loved clothes and spectacular entertainments, and so did her nobles. Under her rule, they became even more powerful than they had been before. They were the owners of land and of people. These people were known as serfs and, like slaves, they had to do whatever their masters wanted. Serfs could be bought and sold by their masters.

Catherine's palaces and the homes of the nobles were filled with furniture, ornaments, carpets, and other luxuries in the latest fashions.

Miserable Existence
Nine out of ten Russians were serfs, and life for them was grim. Often, they lived in poor log huts with only one room for an entire family.

The Hermitage
Catherine built a Winter Palace in St. Petersburg. She called it the Hermitage, because it allowed her to shut herself and her court away like hermits. Today it is a famous museum.

During court entertainments, the wearing of masks became popular.

The court entertainments were dazzling to look at, with gorgeous costumes.

The playing of music was encouraged at court. Catherine invited foreign musicians to Russia to perform their work.

The nobles dressed in clothes that came from fashionable France. They were the finest clothes money could buy.

Revolt of the Serfs
In 1771, a Cossack soldier called Emilian Pugachev set himself up as a rival emperor to Catherine. The Cossacks were a warlike people who lived in southern Russia. Thousands of serfs who wanted to get rid of the nobles joined in their revolt.

The number of rebels under Pugachev grew, and in 1773, they swarmed across Russia, destroying the city of Kazan. Catherine ordered her army to attack the rebels, and in July 1774, the serfs were defeated. Pugachev was captured, brought to Moscow, and executed. The revolt was over.

THE JOURNEYS OF JAMES COOK

In the 18th century, a great explorer called James Cook made three voyages that mapped the Pacific Ocean, the world's largest and deepest sea. On his first voyage, he sailed around New Zealand and down the east coast of Australia. On the second journey, he explored the Antarctic and mapped many South Pacific islands. On his last trip, he discovered Hawaii.

Cook's voyages were important because, unlike many explorers before him, he was not interested in conquest. Instead, he tried to find out about the people, the plants, and the animals of the countries he visited. On board his ship were scientists, artists, and collectors, as well as officers, crew, and servants.

South Sea Paradise
Throughout the 18th century, artists and writers presented life in the South Pacific as being easy and perfect.

Because they were members of the Royal Navy, Cook and his companions wore naval uniforms.

The first voyage to the South
Seas was made to watch the
movement of the planet Venus.
The scientists looked through
a telescope like this.

This very accurate
astronomical clock was used
for timing the observations
of the stars and planets.

Art for All

Drawings and paintings made
by the artists on Cook's voyages were
published. People in Europe could see
illustrations of new plants and animals
discovered on the explorations.

Erythrina

Hibiscus

This is a
quadrant. It
was another
means of
helping sailors
find their way.

Wildcat

Butterfly fish

Blue-crowned lory

Artists were taken on the voyages
to draw the plants and animals
they saw, just as a photographer
might take pictures today.

REVOLUTION

The revolutionaries were against organized religion, and many clergymen fled the country.

In the 18th century, an important revolution started in France. It happened because the king, Louis XVI, and his nobles held all the power and wealth in the country. He could rule the people as he pleased. The way he chose to do so was unfair. For example, the nobles, who were already rich, paid no taxes while the poor peasants did.

In 1789, the king had nearly run out of money. He wanted to increase taxes. At the same time, food was very scarce because of bad harvests. The ordinary people had had enough and decided that they didn't want to be ruled by the king. Louis was executed, and a new government was set up.

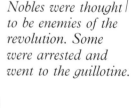

Nobles were thought to be enemies of the revolution. Some were arrested and went to the guillotine.

The revolution was supported by people who thought they had been badly treated by their rulers. They included lawyers, traders, small farmers, workers, and ordinary soldiers.

The nobles made fun of the simple clothes of the revolutionaries. They called them sans-culottes, people without the fine knee-length breeches the nobles wore.

Women played an important part in revolutionary events. They led many of the marches.

The Guillotine

Anyone who was not loyal to the revolution faced arrest and possible death by guillotine. The guillotine cut off the heads of the victims. It was made of a heavy, sharp blade that fell between two posts. Death by guillotine was very quick. It caused less suffering than other methods of execution.

Victims of the guillotine rode to the place of execution in open carts, called tumbrils.

The revolutionaries wore a red bonnet that looked like a nightcap. It was decorated with a blue and white ribbon.

American Revolution

In the 17th century, people left Europe to live in America. They were ruled from Britain and were called colonists. Eventually, they wanted to rule themselves and fought the British for their independence. The French revolutionaries were encouraged by the American success.

This painting shows colonists fighting the British in Massachusetts in 1775.

RICHES OF INDUSTRY

One of the biggest changes in the history of the world, the Industrial Revolution, started in Britain in the late 18th century. As the "Workshop of the World," Britain was the first home of new machines, new types of materials, and new ways of making power. This was the age of coal and iron, of gas and electricity, of railways and factories.

Within 50 years, this series of mighty inventions had dramatically changed the way in which people lived. Railways and steamships made it possible to travel quickly from place to place. Instead of living in the country, many more people lived in towns and cities. There they worked in factories where machines made things in vast numbers, quickly and cheaply.

Modern photography was invented by the British scientist William Fox Talbot during the 1830s. It soon became popular for people to have their photograph taken.

The railways and cheaper paper provided many more readers with news of events from all over the world.

Grim Conditions
The big industrial cities were very smoky, and many people were crammed together in badly built houses.

With the arrival of trains, which had to run according to timetables, people began to live their lives by the clock.

Iron Foundry
Abraham Darby replaced charcoal (made from wood) with coke (made from coal) for making a new kind of tough iron.

The invention of electroplating made it possible to coat iron objects with silver. They looked like solid silver but were far cheaper to make.

Made of Metal
Engineers soon made use of iron and steel. In 1789, Abraham Darby III built the first iron bridge across the river Severn. It is now called Ironbridge. The Eiffel Tower was built in France in 1889, and is nearly 330 yards (300 m) high.

Vast numbers of machine-made cups and plates were turned out for everyday use.

The invention of artificial dyes in the 1850s meant that cloth did not fade when it was washed.

The Great Exhibition
In 1851, the Great Exhibition was opened in London inside a huge glass building called the Crystal Palace. The Exhibition displayed all the latest industrial developments.

Factory Life
The first factories were built to contain the heavy machinery needed to produce cotton cloth. Whole families – even young children – kept the machines going night and day.

PIONEERS

During the 19th century, European settlers traveled across America in search of land to farm. They were called pioneers. Some of them traveled in wagon trains so long that they stretched as far as the eye could see. The wagons were packed tight with provisions – food, tools, plows, household goods, and even chamber pots. There was room only for small children, the sick, and some women to ride in the wagons. Everyone else walked alongside.

Tormented by the heat and dust or by gales, rain, and snow, the pioneers trudged across prairies and climbed over mountains. They lived and slept outdoors and were often short of food and water. The pioneers also faced attacks from Native Americans, who resented the Europeans' taking their land from them.

The wagons were pulled by teams of horses, mules, or oxen.

The wheels at the front were made smaller than those at the back so that the wagon could be steered more easily.

The Lure of Gold
Prospectors were people who hunted for gold. They would fill a shallow pan with gravel from a riverbed and wash the stones out of the pan with water. Then they looked for any gold that might have sunk to the bottom.

When gold was discovered in Australia, America, South Africa, and Canada in the 19th century, Europeans flocked to these countries to make their fortune.

Self-assembly

When the pioneers came to set up home, they had to build their own houses. Some made them out of logs, but others used chunks of dry earth cut from the ground.

The Great Trek

In 1835, Dutch settlers in South Africa moved in wagons to new land to escape being ruled by the British. This journey was called the Great Trek. For safety against the Africans whose land they had entered, the settlers would carefully form their wagons into a circle at night.

The top was made of canvas held up by a frame of hoops. It helped keep out rain and dust.

The hoops were made of strong and flexible hickory wood.

The pioneers had to bring all their cooking equipment with them.

The wheels were made of wood, and the rim was covered in iron. Wheels often broke and held up the wagon train.

THE AMERICAN CIVIL WAR

In 1861, a civil war started in America between states in the South that allowed slavery and states in the North that did not. The South tried to break away from the Union to form a separate nation, but the North went to war to prevent this. After four years of bitter fighting, the North won, and the South was forced to return to the Union. But the price paid by both sides was terrible. Some 620,000 soldiers died, over half of them victims of disease, not of battle. When they fell ill or were wounded, their treatment was as likely to kill them as cure them. The field hospitals were filthy and the surgeons often poorly trained.

If they were lucky, soldiers were given drugs to make them unconscious during surgery. But often they had nothing to ease the pain.

Union states

Confederate states

Border states that fought on the Northern side

Areas with the most slaves

Other territories

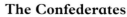

The Confederates
The soldiers of the South, or Confederacy, were given uniforms of gray coats and caps and blue trousers. When uniforms were in short supply, the men wore whatever they could find.

The Unionists

The soldiers of the North, or Union, wore dark blue coats or jackets and sky blue trousers. They had more and better weapons than the troops from the South because most of the factories making weapons were in the North.

The surgeons who treated the soldiers often worked with dirty hands and clothes spattered with blood.

The Slave States

The wealth of the Southern states of America came from plantations producing cotton, sugar, and tobacco. The work on the plantations was carried out by slaves – Africans who had been captured and shipped from Africa over to America. Their masters could do what they liked with them, and they were often ill-treated. Slavery was finally ended in America in 1865.

Slaves could be bought and sold in markets like animals. Often families were split up.

Women volunteered to help the wounded. In the North, teams of trained nurses were set up.

Most medicines were not very effective, and a few were actually dangerous.

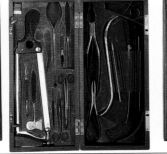

Special Kit

With its pliers and saw, this box looks like a tool kit. In fact, it was a surgeon's case used during the American Civil War.

IN THE SCHOOLROOM

The chalkboard was double-sided. It was on wheels so it could be moved easily.

Until 1870, children in England did not have to go to school. Children from rich families went to school, but poor families could not afford to pay for education. Poor children had to go to work to help their parents. After 1870, the government made school places available to young children for a very small payment. By 1902, education was free, and every child aged between 5 and 13 had to attend school.

The globe of the world was used to teach geography to the children.

At school, children were taught reading, writing, arithmetic, and religion. The children sat at their desks, chanting spellings and tables over and over again and copying words onto slates. At other times, they did some geography and history, drawing, singing, and physical exercise. Discipline was very strict, and children were beaten if they made mistakes in their work.

A wooden hoop used for playing

Playtime

When they were not in school, children amused themselves with outdoor games like the ones played today – marbles, skipping, hopscotch, and football. Hoops were popular, too – they were rolled along the ground, thrown in the air, or whirled around the body.

China inkwells in a tray were filled and given out to the older children by a monitor. The children wrote with pen and ink in a special copybook.

The ink to fill the inkwells was kept in a special container that looked like a small watering can.

The window was built high up so that the children could not look out and be distracted from their work.

Teachers often carried a cane with them. If the children were bad, they would be smacked.

This is an abacus, or counting frame. Children learned to add and subtract by moving the beads along the wires.

Each child in the class had only one reading book. It contained stories and poems and had to last the whole year.

Young children did their writing on slates, using slate pencils. The slates were cheap and could be used over and over again.

The desks were very simple. The shelf underneath held the children's books.

Pencils were made of a single piece of lead, without a wooden casing.

The desks had special holes made in them to hold the inkwells.

Stitch by Stitch
Samplers were very popular in schools. They consisted of a piece of embroidery designed to show a girl's skill in using different stitches. Girls began to make them at an early age. They were a way of teaching girls their letters and their sewing at the same time.

ARTS AND ENTERTAINMENT

Today we are very lucky because we have lots of interesting and entertaining things to see and do. At one time, though, people had to put all their time and energy into finding food, keeping warm, and protecting themselves from attack. It was not until they solved these problems that they began to look for ways of having fun. The earliest forms of entertainment were drawing pictures and telling stories. Later, men and women made simple music and sometimes danced to it.

Now, machines do most of our work for us, and we have time to create and enjoy exciting plays, films, music, ballets, and amazing photographs, paintings, and sculptures.

Recorder   Violin   Trumpet

Swan Queen in the ballet *Swan Lake*

Camera

Powder pigments

Artists' brushes

The Secret,
a sculpture by Auguste Rodin

THEATERS

People have enjoyed going to the theater to watch plays for centuries. The first theaters were built by the Greeks about 2,500 years ago. Stone seats were carved into a hillside.

The actors did not need microphones, because sound carried right up to the back row. Until the 17th century, most plays were staged in the open air. But by the 1650s, bare stages had been replaced by elaborate sets that had to be kept inside, so indoor theaters became common. Lights transformed theaters, too. Realistic acting began with the invention of limelight – for the first time, players' expressions could be seen.

Wagons Roll
In the Middle Ages, actors took their shows, called miracle plays, from town to town. These plays were about the battle between good and evil.

The upper gallery was about 26 feet (8 meters) above the ground.

Stage door

The Globe Theater was built in 1599.

More than 2,000 people could crowd in to watch a good play. Only a few could afford to sit in these galleries.

Ancient Art
Noh is an old, traditional type of Japanese theater. Religious stories and ancient myths are performed on a stage that has very little scenery. The plays, some of which are more than 500 years old, can go on for as long as six hours!

The Globe Theater

William Shakespeare, the most famous of all playwrights, acted on the Globe's wooden stage.

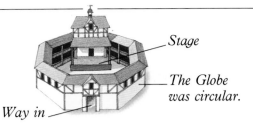

Stage

The Globe was circular.

Way in

Musicians played on this balcony.

When the flag was raised, people knew that a play was going to be performed.

There was no scenery on the stage.

This model of the Globe Theater has been cut in half to let you see inside.

Most of the audience stood around the stage. These "groundlings" got wet when it rained.

Setting the Scene

Theaters come in all sorts of shapes and sizes. Early ones were open to the air. Most modern stages have roofs so that plays can be staged when it is raining.

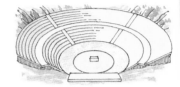

Greek theater
200 BC

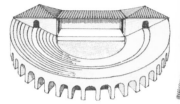

Roman theater
AD 100

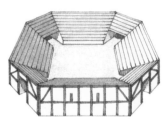

Elizabethan theater
16th century

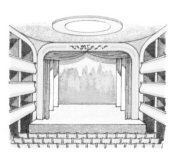

"Modern" theater
19th century

PLAYS AND PLAYERS

Playwrights write stories, or plays, that are performed on stage by players, called actors and actresses. Molière and Shakespeare both lived more than 300 years ago, but their plays, such as *L'Avare* and *Macbeth*, are still popular. Laurence Olivier, one of this century's most famous actors, starred in *Henry V* by Shakespeare and also in more modern plays, such as *The Entertainer* by John Osborne. Like all good players, he could make an audience believe that what they are seeing is real, not just an act.

WILLIAM SHAKESPEARE

Not a Word
Marcel Marceau is a famous French mime artist – he acts out stories without speaking.

The actor playing Romeo shows that he likes Juliet by giving her a rose.

The players wear costumes like those worn by rich Italians in the past.

Happy Ending
Plays that make people laugh are called comedies. Molière wrote many wonderful plays in the 17th century that are still funny today.

Sad Ending
Tragedies are plays with sad endings. *Cat on a Hot Tin Roof* was written by Tennessee Williams and is a famous modern tragedy.

Setting the Scene

Romeo and Juliet is one of Shakespeare's saddest plays. The couple can never be together because their families hate one another.

These are the first words Romeo ever says to Juliet.

ROMEO (*to Juliet, touching her hand*)
If I profane with my unworthiest hand
This holy shrine, the gentle sin is this:
My lips, two blushing pilgrims, ready stand
   To smooth that rough touch with a
      tender kiss.   95

JULIET
Good pilgrim, you do wrong your hand
      too much,
Which mannerly devotion shows in this;
For saints have hands that pilgrims'
      hands do touch,

Juliet replies. She is not upset with him for holding her hand.

This line is not spoken – it tells the actors what to do.

There are 145 lines in this scene, or part of the play. This is line 95.

Makeup Magic

To make an audience believe what they are seeing, players often change the way they look by using makeup.

Young

Going gray

Talcum powder and special white mascara makes brown hair and eyebrows look gray.

Pale powder takes away the rosy cheeks.

Lines are drawn on the face with a special dark pencil.

Old

The actress pretends to be shy, yet pleased to receive the rose.

Staged in the East

Spectacular, traditional Chinese plays are known as Peking opera in the West. Every action and color has a special meaning. Blue and yellow costumes show that a person is bad tempered and proud.

MUSIC

Musical sounds can be made with your voice or by playing an instrument. The first instruments were played more than 35,000 years ago – people blew into shells and hollow mammoth bones! Today's instruments are more complicated. The sounds they make can be high or low. This is known as the pitch of the note. The way these sounds are arranged is called the tune. Rhythm is the pattern of long and short notes. A skillful musician can make the same tune sound slow and sleepy or loud and jazzy!

Beat the Clock?
A metronome can tick at different speed Musicians listen to it to make sure they are playing at the right speed. This is called keeping the beat.

Stylish Groups

Jazz band

Pop group

String quartet

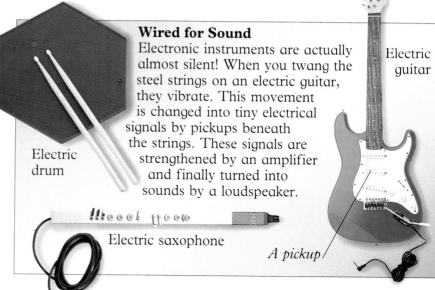

Wired for Sound
Electronic instruments are actually almost silent! When you twang the steel strings on an electric guitar, they vibrate. This movement is changed into tiny electrical signals by pickups beneath the strings. These signals are strengthened by an amplifier and finally turned into sounds by a loudspeaker.

Electric drum

Electric saxophone

Electric guitar

A pickup

Note it Down
Music is written down in a special language. Instead of words, there are notes. These are the notes for "Here Comes the Bride."

Deeper notes are written on lower lines or in lower spaces.

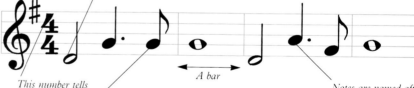

This number tells you there are four beats in each bar.

A bar

The way a note looks indicates how long it lasts. Notes like this are the shortest in this tune.

Notes are named after letters. This one in the second space up is A.

Bagpipes

You blow here.

Scotland is famous for bagpipes. They are played by blowing air into a bag and then squeezing the air up through the pipes.

A World of Music

Different countries and regions have very different instruments and styles of music. This traditional music is played by local people, or folk, so it is often called folk music.

Banjo

People who live in the eastern mountains of the United States are famous for their banjo playing. Banjos were first brought to the United States by enslaved Africans.

Panpipes are made from pipes of different lengths and are popular in South America. You play them by blowing over the top of the pipes. Longer pipes make lower sounds.

Panpipes

Australian aboriginals are the only people who play the didgeridoo. It is usually made from a hollow bamboo branch and is difficult to play.

Didgeridoo

ORCHESTRAS

Many musicians play music in groups, called orchestras. Most orchestras have four sections: string, percussion, woodwind, and brass. The different sounds and notes, from as many as 120 instruments in a symphony orchestra, combine to form marvelous music. Orchestras usually play classical music, often written by great composers of the past, such as Mozart and Beethoven.

Percussion players play many different instruments, such as drums, triangles, and gongs.

Cymbal

Principal violinist

Clarinet

Violin

Sounds Different

Most of the instruments in a gamelan orchestra are percussion instruments – there are no violins at all. This music comes from Indonesia.

The leader of an orchestra is always a violin player who sits near the conductor.

Woodwind instruments, including recorders, are played by blowing down a tube that has holes in it.

Two beats to a bar

Three beats to a bar

Four beats to a bar

Keeping the Beat

If all the instruments in an orchestra played at different speeds, they would sound dreadful. The musicians keep in time by watching a conductor, who waves a baton to the beat of the music.

Conductor

Short stick, called a baton

The conductor stands in front of the musicians so that they can see the baton.

The French horn, like the trumpet and trombone, is a long, curving tube that is made of brass.

Bow

This cello, like the violin, is part of the string section. It is played by drawing a bow across the strings.

Which Section?

The clarinet player is sitting on a yellow seat. By using the key shown below, you can see that a clarinet is a woodwind instrument.

Strings

	String		Percussion		Woodwind		Brass

Child Genius

Wolfgang Amadeus Mozart wrote, or composed, over 600 pieces of music. He completed his first symphony when he was eight years old!

DANCE

Dancing is a way of moving your body in time to music. Early peoples danced to please their gods. Today, most people dance purely for fun. It was not until the 12th century that dancing in pairs became popular. The arrival of the waltz in the 1800s was shocking – it was the first time that couples had held one another closely as they danced. A revolution has taken place this century, too – set steps have been replaced by solo dancers who make up all their own moves.

This is an imaginary "bow."

The dancer pulls back an "arrow."

There are more than 25 different finger positions. Each one has a different meaning.

Proud to Dance
The spectacular flamenco dance comes from the gypsy peoples of Spain. The dancers hold their heads up high as they stamp their feet and turn around. The women wear long, colorful dresses, but the men wear black.

Head over Heels
Jive dancing has few rules. The dancers twist, spin, jump, leap around – whatever the music makes them feel like doing!

Taking a bow

The dance begins.

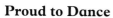

Taps are fitted to the heel and toe of each shoe.

Tap, Tap, Tap

Fred Astaire was one of the most famous dancers of this century. He tap-danced his way through many shows and movies. Small pieces of metal, called taps, were fixed to the soles of his shoes. When his feet touched the ground, you could hear the quick clatter of clicking metal.

A classical Indian dancer uses the whole of the body. The neck, wrists, and even the eyes move to the rhythm of the music.

As she pretends to put a clip in her hair, the dancer looks into the "mirror."

Her hand is held out flat to form a "mirror!"

Moves Matter

Classical Indian dance has been performed for more than 1,000 years. Sometimes the dancers just create shapes and patterns with their bodies. In other dances, they use hand movements and mime to tell stories of Hindu gods.

She stamps her bare foot in time to the drumbeat of an instrument, called a tabla.

The pleats in the special dance dress let the dancer move freely.

One, Two, Three

The waltz is danced in triple time – each bar of music has three beats. As the dancers swirl around and around, they draw their feet together once every three steps.

BALLET

Ballet is a graceful type of dance that uses particular positions and steps. It began in Italy, but was developed by the French into the style you see today. Louis XIV of France started the first ballet school in 1661. But the dancers in his "ballets" at the French court sang and recited poems! The first true ballet, without words, was not performed until 1789. The basic steps and jumps taught by Louis's school are still used today, which is why many of them have French names. *Glisser* means "to glide" and *pas de chat* means "cat step"!

At the Bar
Ballet dancers must warm up their muscles before they dance. They do this by stretching while holding onto a pole, called a bar.

Hopping Frogs
The Tales of Beatrix Potter is an amazing ballet. The dancers wear wonderful costumes and animal masks.

This short, stiff skirt is called a tutu. It allows the dancer's legs to be seen.

Women did not dance in ballets until 1681. Then, they had to wear long, flowing dresses.

"Turning out" legs and feet takes years of practice.

First position
(*en première*)

Second position
(*en seconde*)

Third position
(*en troisième*)

Fourth position
(*en quatrième*)

Fifth position
(*en cinquième*)

The Five Positions
In the 17th century, a ballet teacher called Beauchamps developed five foot and arm positions that enabled dancers to keep their balance and still look elegant. Most ballet steps begin or end with one of these positions.

Long hair is always tied back in a bun.

Male dancers need to be strong. They have to lift ballerinas high up in the air.

This Is Ballet, Too!

During this century, some very adventurous ballets have been performed. These modern ballets use many of the classic steps, but often have no story. The imaginatively dressed dancers simply express a mood.

Ballerinas sew on their own ribbons.

The ballerina keeps her foot pointed.

The leg is held very still and straight.

A female ballet dancer is called a ballerina.

Dancing on tiptoes makes the ballerina's legs look longer and more graceful.

Ballerinas put cotton inside their shoes to keep their toes from hurting.

Ballet dancers wear tights.

Ballet Shoes

To dance on tiptoe, or *en pointe*, ballerinas wear special satin slippers with stiffened toes. Girls have to be at least twelve years old before their feet are strong enough to dance like this.

Male dancers' shoes do not have stiffened toes.

OPERA

Inside Paris opera house

Operas mix music with theater in a very spectacular way. The performers on stage act out a story, but instead of speaking the words, they sing them. A performance usually begins with an overture. This is a piece of catchy music that features snatches of the tunes that will be heard during the rest of the opera. As well as singing the story, the stars on the stage also perform solo songs, called arias. These are often the most beautiful songs.

Smashing Note

When you rub your finger around the rim of a glass, it makes a high sound. If a person with a strong voice sings this high note loudly, and for a long time, the sound waves can shatter the glass!

Stylish Surroundings

Operas are usually staged in special buildings that are designed to help the whole audience hear the sound clearly. The Sydney Opera House in Australia is one of the most modern. It was finished in 1973.

Paris opera house

The jagged roof looks like waves or the sails on a yacht.

Just one of the two main halls holds 1,500 people.

Sydney Opera House

Crowded Stage

Operas sometimes have spectacular stage sets and, as well as the soloists, a large group of singers, called the chorus.

A Choice of Voices

The soprano is the highest female voice. It is a brilliant and exciting sound.

The contralto, or alto, has a slightly lower voice than the soprano. It has a warmer tone.

The tenors sing many of the important male songs in an opera.

The bass sings the lowest notes, and his voice is deep and strong.

Dressed to Impress

Opera singers wear incredible costumes to play their parts. In *The Cunning Little Vixen*, their makeup is amazing, too.

On Song

Luciano Pavarotti's voice is powerful and beautiful. He is one of the world's greatest tenors.

Operas are mostly sung in Italian, French, or German.

Powerful lungs enable the singers to sing without a microphone!

Foxy Tale

Operas tell stories. *The Cunning Little Vixen* by Leoš Janáček is about a crafty female fox who escapes from a forester and has many exciting adventures.

PAINTING

This says "Jan van Eyck was here, 1434." It may be to prove the artist was a witness to the marriage.

This cave painting is believed to be 30,000 years old.

Many thousands of years ago, people painted pictures of bulls, horses, and antelopes on the walls of their caves. We will never be sure why they did this – perhaps it was to make magic and bring people luck in their hunting. Since then, artists have painted pictures to record events, to honor heroes and heroines, to make people wonder about the world, to tell stories, or simply for pleasure. All artists have their own painting styles, and you may not like them all – it's up to you.

If you look closely in the mirror, you can see the reflections of two other people. Perhaps one is the artist himself.

Hills and Valleys
Paintings of the countryside are called landscapes. This Chinese scene looks very different from a Western landscape. It is painted on silk, using a fine and detailed style.

Things that are farther away look smaller, so the artist has made the shoes at the front bigger than those at the back. This gives the picture depth.

King Ramses II

by an unknown Egyptian artist
(detail)

Lucrezia Panciatichi

by Il Bronzino
(detail)

Self-portrait

by Vincent van Gogh
(detail)

Woman Weeping

by Pablo Picasso
(detail)

The Arnolfini Marriage, painted by Jan van Eyck in 1434

Looking for Clues
This painting records a couple's marriage. But if you look closely, you will see it is much more than just a wedding picture.

One candle is left to burn, even though it is daylight, to show that the couple will always love one another.

Like the merchant, his wife is wearing expensive clothes. It was the fashion to hold skirts in this way.

Action Painting
In this painting, the modern American artist, Jackson Pollock, was not trying to show objects or people, but feelings. He worked by dripping and splattering paint on the canvas.

The light shines on the couple's hands, and the artist has shown every tiny line.

Painting Portraits
For thousands of years, artists have enjoyed painting pictures of people. Not all of them painted the person true to life – there are many different styles of painting.

Son of Man

by René Magritte
(detail)

Marilyn Monroe

by Andy Warhol
(detail)

SCULPTURE

Sculptures are three-dimensional, not flat like paintings. Bricks, plastic, even rubber tires have all been used by sculptors, but the traditional materials are wood, stone, and clay. Some sculptors build up their image by adding small pieces of material, such as clay. Others start with a big block of wood or stone and cut away, or subtract, material. When Michelangelo carved, he believed he was setting free a stone person, or statue, that was trapped inside the stone!

Mystery from the Past
Strange statues were carved on Easter Island more than 1,000 years ago. Nobody knows for sure what they were for, or how they were built.

Many of Moore's sculptures are based on natural shapes, such as smooth pebbles.

In Touch with Nature
Some of the sculptures created by Henry Moore (1898-1986) look like everyday things, but many are just interesting shapes.

This rounded sculpture is meant to be touched.

If you walk around this sculpture, you see different, exciting shapes.

Mountain Men
Mount Rushmore has the heads of four American presidents cut into it. It took 14 years to blast and drill presidents Washington, Jefferson, Roosevelt, and Lincoln out of the rock.

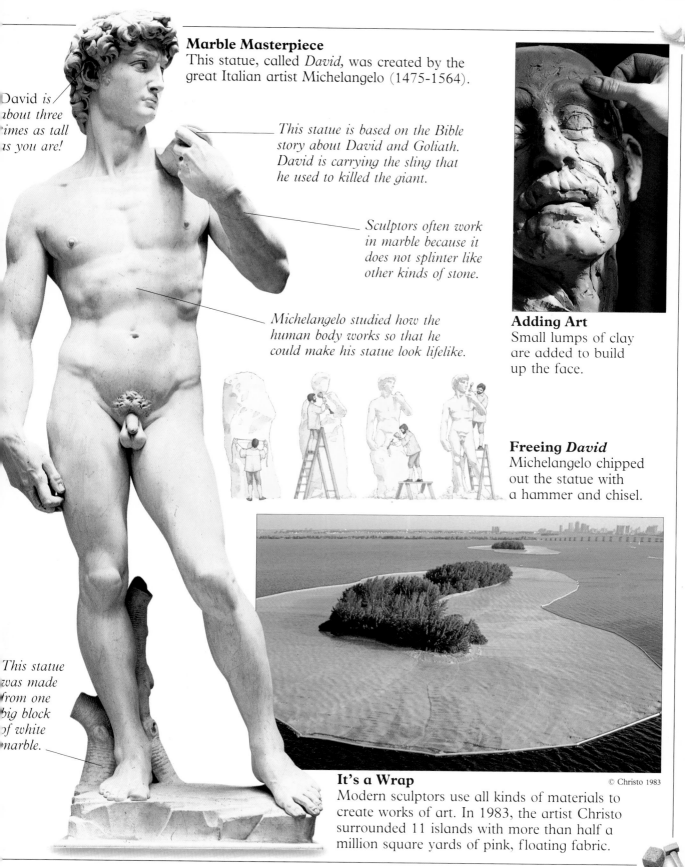

Marble Masterpiece
This statue, called *David*, was created by the great Italian artist Michelangelo (1475-1564).

David is about three times as tall as you are!

This statue is based on the Bible story about David and Goliath. David is carrying the sling that he used to killed the giant.

Sculptors often work in marble because it does not splinter like other kinds of stone.

Michelangelo studied how the human body works so that he could make his statue look lifelike.

This statue was made from one big block of white marble.

Adding Art
Small lumps of clay are added to build up the face.

Freeing *David*
Michelangelo chipped out the statue with a hammer and chisel.

© Christo 1983

It's a Wrap
Modern sculptors use all kinds of materials to create works of art. In 1983, the artist Christo surrounded 11 islands with more than half a million square yards of pink, floating fabric.

PHOTOGRAPHY

Every day, more than 2 million photographs are taken around the world. These photos are not just snapshots of birthdays and holidays, but also pictures of goals being scored, models in studios, and important events. No matter how complicated the camera, all photos are made in much the same way as when the process was invented more than 150 years ago. A camera is basically a box with a hole in it. Light enters through this hole and shines onto the light-sensitive paper inside to form a picture.

Say Cheese!
In early photographs, it took a long time for the picture to form. People had to stay still for up to 20 minutes to keep the picture from blurring

Flash

This tough camera will not break if it is dropped!

This spare camera has a long "telephoto" lens. It makes things that are a long way off look closer.

Lenses and lots of rolls of film are stored in this bag.

Click! A sinking bus is snapped.

Right Time, Right Place
Newspaper photographers have to be ready for action all the time. They may not get a second chance to snap an amazing event or famous person.

Camera Collection

35mm professional camera

Instant camera

High-quality studio camera

Flash

Underwater camera

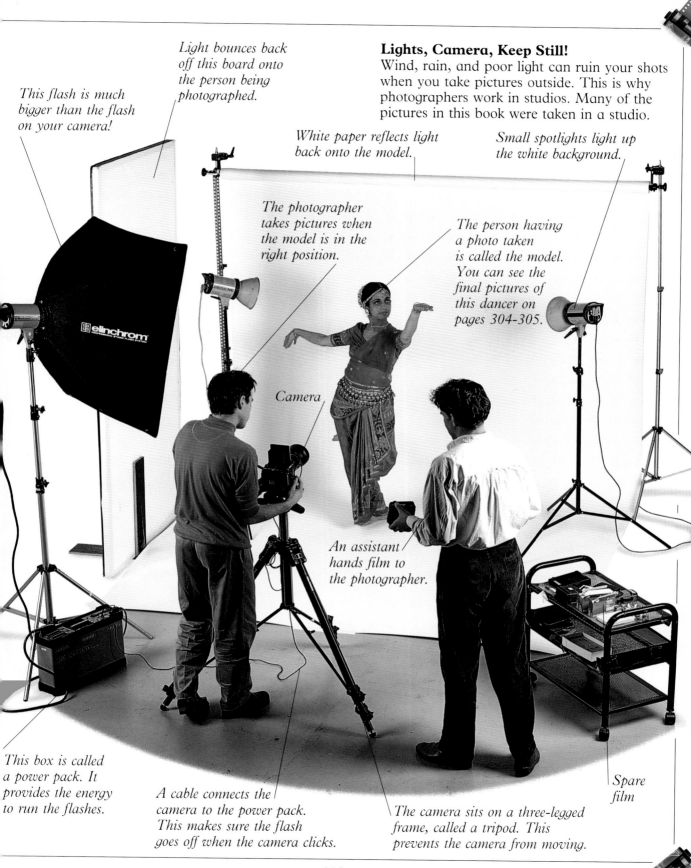

This flash is much bigger than the flash on your camera!

Light bounces back off this board onto the person being photographed.

Lights, Camera, Keep Still!
Wind, rain, and poor light can ruin your shots when you take pictures outside. This is why photographers work in studios. Many of the pictures in this book were taken in a studio.

White paper reflects light back onto the model.

Small spotlights light up the white background.

The photographer takes pictures when the model is in the right position.

The person having a photo taken is called the model. You can see the final pictures of this dancer on pages 304-305.

Camera

An assistant hands film to the photographer.

This box is called a power pack. It provides the energy to run the flashes.

A cable connects the camera to the power pack. This makes sure the flash goes off when the camera clicks.

The camera sits on a three-legged frame, called a tripod. This prevents the camera from moving.

Spare film

MOVIES

From the first flickery film, people have loved the world of movie make-believe. The early films were made in black and white and did not have any sound. They were called silent movies. In the theater, a pianist played along with the film, and the audience had to read words on the screen. It was not until 1927 that the first "talkie," called *The Jazz Singer*, came out. Color films, such as *The Wizard of Oz*, were made in 1939, but only became common in the 1950s.

A Star Is Born
The dry weather in California is ideal for filming. Hollywood, once a sleepy suburb of Los Angeles, was the center of the American film industry by 1920

Cue Clapper Board
When a clapper board is clicked shut, it is seen on the film and heard on the soundtrack. The pictures and words can then be matched.

Silent Star
Charlie Chaplin (1889-1977) was one of the first movie stars. He starred in more than 80 movies. His most famous character was a little tramp who had a funny walk.

Moving Pictures
A movie is made of many still pictures, called frames. The frames pass before your eyes so quickly that the pictures look as if they are moving.

Small holes keep the film steady in the camera.

Frame

Eastern Promise
More than 800 movies a year are made in India – double the number made in Hollywood. Streets in Bombay and Delhi are lined with bright posters that advertise these dramatic movies.

The lighting crew sets up the lights so that the actor looks as if he is outside and not inside a bright studio.

This "mountain" was built inside a studio.

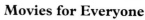

Rain machine

The camera rolls forward on a wheeled truck, called a dolly.

Romance

Western

Horror

Science fiction

Between shots, the makeup artists make sure that the actor looks right.

This pole, which has a microphone on the end of it, is called a boom. It is held up high so that it does not appear in the movie.

The director is in charge of shooting the movie.

SPECIAL EFFECTS

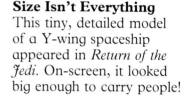

Have you ever wondered how alien worlds are filmed or how actors and actresses survive death-defying leaps? These amazing events are not real. They are created by special effects. With clever camera work and special equipment, such as wind machines, film directors can make us believe that what we are seeing is real. These tricks of the trade are used when shooting the action in the normal way would be too expensive, too dangerous, or just impossible!

Size Isn't Everything
This tiny, detailed model of a Y-wing spaceship appeared in *Return of the Jedi*. On-screen, it looked big enough to carry people!

A wind machine can make a gentle breeze or a howling gale.

Cue Weather
Filmmakers can't wait for the right weather. In this scene from *The Mosquito Coast*, giant fans whipped up the waves that battered the boat.

The lights shining through the glass look like real camp fires.

The special-effects expert pulls the trigger to fire the red pellet.

Blood Gun

People often have to pretend to be shot in films. Real guns can't be used because this would be too dangerous, so special blood guns are used. A small pellet is filled with red liquid. When this hits bare skin, it bursts and looks like a nasty bullet wound.

Air is forced down this hose to push out the blood bullet.

Land of Make-Believe

This treetop village appeared in the movie *Return of the Jedi*. It was not built in trees, but painted onto glass! Small parts of the painting are scraped away, and live action is shown through these holes. In the finished film, actors, dressed as alien Ewoks, look as if they are in the village!

This special trick painting and filming is called matte painting.

Fire!

Filmmakers don't just throw bombs to make flames and smoke. Explosions are carefully set up and controlled.

Fuse

Smoke cylinders

Explosives

This can makes orange smoke.

The highly trained stuntman leaps from the building.

The audience must think that the actor is falling, so the stuntman wears the same clothes.

The stuntman waves his legs and arms around, to look as though he is terrified.

The camera stops running when the stuntman prepares to land.

Falling for You

Jumping from a high building is dangerous and difficult! In movies, it is only done by experts, called stunt people. They take the place of the actor or actress who is supposed to be falling to the ground in the story.

The stuntman twists in the air so that his shoulders will hit the bag first.

Actor

The stuntman hits the soft air bag at more than 50 miles (80 km) per hour!

FOOD AND FARMING

Most of us buy our food in stores or from market stalls, but it has probably been produced somewhere far away. This could be a large farm in the country or a small plantation halfway around the world.

We eat some things – fruit, raw vegetables, and nuts, for example – just as they are picked. Other types of food have lots of things done to them before we put them on the table. Cocoa beans, for example, which grow in huge pods on cacao trees, are made into delicious chocolate. Grains of wheat are threshed to separate them from their husks, then milled into flour. Flour is used to make the bread, biscuits, and pastry that most people eat every day.

Chocolate cake

Baby pigs are called piglets.

Baby sheep are called lambs.

Fruits and vegetables

Many different kinds of cheese are made from milk.

Broccoli farming

Wheat and bread

Tractor

WHEAT

What do breakfast cereals, breads, cakes, pastries, pasta, and couscous have in common? The answer is that they are made out of wheat, the most important crop in the world. Wheat is the main food eaten every day by 35 percent of the world's people. It is so important, it is called a staple food. Like barley, oats, rye, corn, rice, sorghum, and millet, wheat is a cereal. Cereals are grasses, and we eat their seeds.

Crop Rotation

If the same crop is grown on the same land year after year, the soil loses its fertility, and pests and diseases build up. The farmer can prevent this by changing the crops grown each year. In the first year, rapeseed may be grown, then wheat, fava beans, wheat again, and barley.

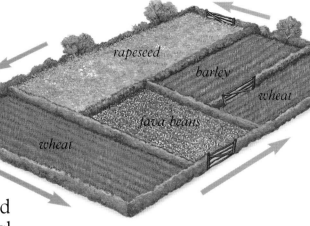

rapeseed

barley

wheat

fava beans

wheat

Machine Power

This is a combine harvester. It cuts the crop and separates the grain from the straw.

This is the unloading spout. It is used to empty threshed grain from the harvester.

In the harvester, the grain is shaken off the stalks. The stalks fall to the ground at the back.

These are straw walkers. They carry the crop through the machine.

Edible Ears

Other cereal crops include oats, barley, and rye. Like wheat, these plants have an "ear" of grain on each stem. Each grain is protected by a husk, and inside the husk is the seed. The grain can be eaten whole or ground into flour.

Rye

Oats

Barley

Wheat Products

Wheat can be used for many different foods. You would never guess that these items are all made from the same ingredient.

Wheat germ

Cracked wheat

Breakfast cereal

Pasta

Monsters on the Move

These wheat fields in the prairies of Manitoba, Canada, are so vast that several combine harvesters can work the fields at the same time.

The cab is air-conditioned to protect the driver from the heat and the dust.

An ear of wheat is made up of 40 to 60 grains. When wheat is ripe and ready for harvesting, it turns golden yellow.

Home Sweet Home

The wheat field is a world of its own. It is home to mammals, insects, birds, and wildflowers.

The reel sweeps the crop into the cutter bar. The cutter bar cuts the crop.

CORN

In many tropical and subtropical countries, corn, or maize, is the main food that people eat. It is eaten as a vegetable or ground into flour or cornmeal. Corn is also made into oil for cooking and for salad dressings. Corn needs lots of sunshine to grow. It will grow in hot climates and also in cooler, mild ones as long as there is plenty of sun. However, the varieties of corn grown in these different climates are not the same.

The flower at the top of the plant is called the tassel.

This is the stalk. Some kinds of corn have stalks as high as 20 feet (6 meters).

Corn is a huge grass.

These long, soft threads are called the silk.

The corn ear, or cob, is protected by a husk of tightly wrapped leaves.

The cob is covered in neat, straight rows of kernels. The kernels are the plant's seeds and the part you eat.

Pest Problems
In warm and tropical areas, one of the pests most feared by farmers is the locust. Locusts travel in huge swarms of over 50 billion insects, stripping plants of their leaves and stems as they feed. Locusts can destroy crops and cause famine and starvation.

Short side roots spread over the surface of the soil. They anchor the tall plant like the guy ropes of a tent and keep it from falling over.

Deep roots go down into the soil and take out the food and water the plant needs to grow.

Sun Ripened

Corn is the staple food of the warmer parts of North, Central, and South America, and also of Africa. It can be grown in huge fields with the help of machinery or on small plots where the crop is tended by hand.

Sweet corn

Breakfast cereal

Mountain of Corn

After the corn has been harvested, it is sorted. Some will be kept as seed for the next crop, some kept for food, and some used as animal feed.

Tortilla chips

Cornucopia of Color

Corn comes in all sorts of colors. It may even be striped, streaked, speckled, or spotted.

Taco

Corn oil

Tough Grains

In dry parts of Africa and Asia, people eat mainly sorghum and millet. These cereals are tough and will grow with little water. This woman is pounding sorghum grain. It is mixed with water or milk until it becomes soft and sticky, like farina. Sorghum is eaten with spicy stews, which give it flavor. It is also ground into flour. Millet is cooked like oatmeal, baked into a flat bread, or used in soups and stews.

RICE

Like corn and wheat, rice is a staple food. It is the main food of over half the world's population – around 2.5 billion people. Ninety percent of the world's rice is grown in Asia. However, the Asian countries eat most of what they grow. The United States also produces huge amounts of rice, but sells much of what it grows to the rest of the world.

Rice is a swamp plant. It grows with its roots in water. It has hollow stems, which take oxygen to the roots.

In Asia, the work of sowing, planting the seedlings, and harvesting the rice is usually done by hand.

How the Rice Plant Is Used

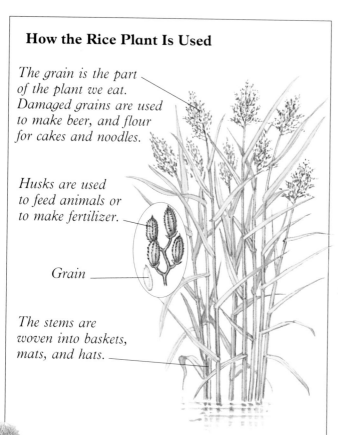

The grain is the part of the plant we eat. Damaged grains are used to make beer, and flour for cakes and noodles.

Husks are used to feed animals or to make fertilizer.

Grain

The stems are woven into baskets, mats, and hats.

Preparing the Fields
The paddy fields are flooded, plowed, raked, and flattened. In Asia, buffaloes are used to do the heavy work.

Planting Out
The seedlings, grown earlier, are moved to the paddy fields. They are planted far apart so they have plenty of space to grow.

While water helps the rice grow, it kills off weeds that can't grow in such wet conditions.

Machine Power

In the United States, growing rice is highly mechanized. Tractors prepare the fields. The seed may be sown from airplanes. Combine harvesters gather the ripe plants.

Rice is a type of grass.

Most rice is grown in flooded fields, called paddies.

The fields are flooded for much of the growing season.

The rice is planted in straight lines.

Safekeeping

Special buildings are used to store the rice until it is needed. They are called granaries. A granary keeps the grain dry and safe from hungry animals.

All Rice

Puffed rice cakes

Rice noodles

Sticky rice cakes

Rice paper

Harvesting

The rice is ready for harvesting in three to six months. The fields are drained and the plants cut down, tied in bundles, and left to dry.

Winnowing

The rice is beaten to remove the grains. The grains are crushed and then sieved and tossed to remove any fine pieces of husk.

VEGETABLE OILS

Vegetable oils are made from the seeds and fruits of many plants growing all over the world, from tiny sesame seeds to big, juicy coconuts. The oils are used for cooking, as salad dressings, and in margarine and cooking fats. Soybeans are the most important source of oil worldwide, especially in the United States. In western Europe, the oil most widely produced is rapeseed oil.

Harvesting

Olives are picked in the autumn when they are ripe. They are shaken from the trees and gathered up, usually by hand. Next, the olives are sorted out to remove the leaves and twigs and taken to be pressed.

Olive trees are often quite small, but can live for hundreds of years. They develop very wrinkled, knotty trunks.

How Olive Oil Is Made

The olives are ground into a thick paste, which is spread onto special mats. The mats are then layered up on the pressing machine, which will gently squeeze them to produce olive oil.

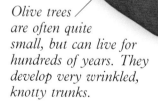

Olive trees are planted in rows. Fields of olive trees are called groves.

Edible Olives

There are many varieties of olives – black and green, large and small. Most are used for making oil, but some are eaten whole, too. Raw olives are very bitter, but once they have been treated, fermented, and pickled, they taste delicious!

Surprising Sources

Oil can be produced from lots of very different fruits and seeds.

Peanuts

Cottonseeds

Sesame seeds

The trees are shaken to make the olives fall. Sometimes, machines are used to do this job.

Each long, flexible branch has lots of flowers and about 30 clusters of fruit.

Rapeseed

Soybeans

Sunflower Oil

Every sunflower is made up of hundreds of tiny flowers surrounded by a fringe of petals. It is the seeds of the flowers that are rich in oil. Sunflowers are ready for harvesting when the flowers are dead and the heads have dried out.

Sunflower oil is good for making cooking oil, salad oil, and margarine.

Nets are laid down so that the fallen fruit can be gathered easily.

CATTLE AND DAIRY FARMING

Cattle are very popular domesticated animals because they provide us with a lot of food. They are kept for their meat, which is called beef, and also for their milk. Milk is full of calcium and protein, which makes it a very complete and nourishing food. Because of this, milk-giving animals are kept all over the world – but they are not always cows. Goats, sheep, camels, reindeer, and llamas give us milk, too!

A cow gives milk for ten months of the year. Then she has a rest for two months. Most cows are milked twice a day.

To produce milk, a dairy cow must have a calf once a year.

Milky Ways
To make just 1 pound (about half a kilogram) of butter, you need a little over 2.5 gallons (9.8 liters) of milk. The leftover liquid is called buttermilk.

A calf born to a dairy cow stays with its mother for just 48 hours.

The Big Cheese
Foods made from milk, like butter, cheese, yogurt, and cream, are known as dairy products. Making cheese is the best way to turn milk into a food that can be stored for some time. Thousands of different cheeses are made all over the world. Soft, creamy cheeses must be eaten promptly, but hard, crumbly cheeses will last for months and even years!

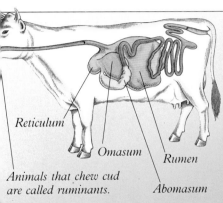

Second Helping

Cows eat grass without chewing it properly. It goes to the rumen and reticulum to be broken down into cud. The cow sucks this up and chews it again. When it is swallowed, the cud goes through the stomachs in turn, finally being digested in the abomasum.

Reticulum

Omasum

Rumen

Abomasum

Animals that chew cud are called ruminants.

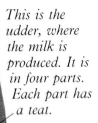

This is the udder, where the milk is produced. It is in four parts. Each part has a teat.

In the Milking Parlor

When cows are milked, special suckers are attached to the teats. They squeeze the milk gently from the cow just as a calf does. Most cows give about 3 gallons (11.5 liters) of milk per day.

Bred for Beef

Big, heavy breeds of cattle are kept for beef because they have more meat on them. The world's largest herds of beef cattle are found in North and South America. Here there are lots of wide, open spaces where the cattle can graze until they are fat and ready for eating.

Dairy Cattle

Holstein-friesian

Jersey

Beef Cattle

Hereford

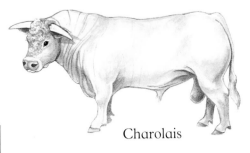

Charolais

Dairy and Beef

Simmental

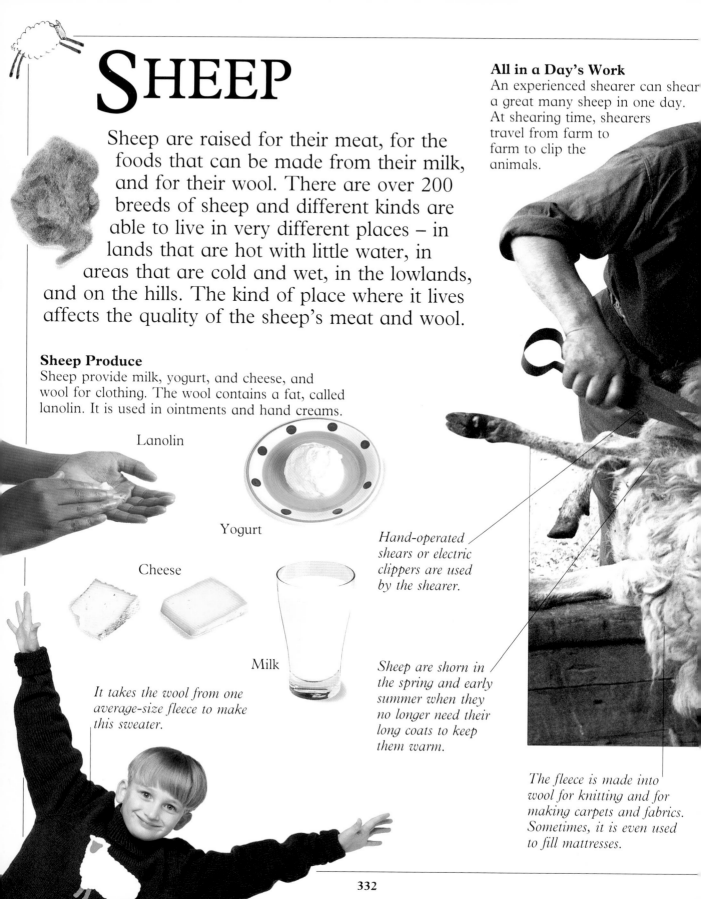

SHEEP

Sheep are raised for their meat, for the foods that can be made from their milk, and for their wool. There are over 200 breeds of sheep and different kinds are able to live in very different places – in lands that are hot with little water, in areas that are cold and wet, in the lowlands, and on the hills. The kind of place where it lives affects the quality of the sheep's meat and wool.

All in a Day's Work
An experienced shearer can shear a great many sheep in one day. At shearing time, shearers travel from farm to farm to clip the animals.

Sheep Produce

Sheep provide milk, yogurt, and cheese, and wool for clothing. The wool contains a fat, called lanolin. It is used in ointments and hand creams.

Lanolin

Yogurt

Cheese

Milk

Hand-operated shears or electric clippers are used by the shearer.

Sheep are shorn in the spring and early summer when they no longer need their long coats to keep them warm.

It takes the wool from one average-size fleece to make this sweater.

The fleece is made into wool for knitting and for making carpets and fabrics. Sometimes, it is even used to fill mattresses.

Bath Time

Once or twice a year, the farmer drives the sheep through a chemical dip. This keeps them free of pests and diseases.

Suffolk

Mommy!

Some lambs are orphaned or not wanted by their mom. These may be bottle-fed by the farmer or shepherd or adopted by another ewe.

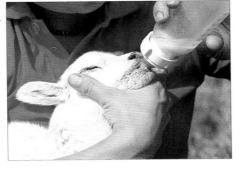

Wensleydale

Working Dogs

Dogs are used to gather the sheep in flocks and move them from place to place. Sheepdogs naturally like to herd animals, but they need to be trained to follow instructions. The shepherd controls the dog using special calls and whistles. Sheep are timid, so the dog will run low in the grass to avoid frightening them.

Karakul

Scottish blackface

In a Spin

To produce wool for knitting or weaving, many strands are twisted together to make one long thread. Spinning was first done by hand using a spindle. Later, the spindle was attached to a wheel, and wool could be spun more quickly. Now, most spinning is done by machine.

Wild Sheep

Bighorn

The shearer shears the sheep so that the fleece comes off in one piece.

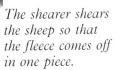

PIGS

Duroc

Domestic pigs are mainly kept for their meat, especially in China, which has the most pigs in the world. Pigs raised for their meat are usually kept in pens and are fed on cereals, potatoes, fish meal, and skimmed milk. Pigs kept for breeding often live outdoors in fields. The farmer feeds them, but they also search for their own food. They like worms, snails, roots, and plants.

A female pig is called a sow.

Mud, Glorious Mud
Despite their reputation, pigs are not dirty animals. They cannot sweat, so in hot weather they roll about in mud to help cool themselves down.

The piglets go to the same teats every time they want to feed. When they are six weeks old, they move on from milk to solid food.

Intensive Farming
When pigs leave their mothers, they are normally housed in piggeries. This is so they can be looked after properly and kept warm. The pigs are fed a special diet, and because they have less space to run around, they put on weight quickly.

A sow makes a "nest" before she has her litter. In a pen, she paws straw into a heap. In a field, she lines a hole with leaves and straw.

Landrace

British saddleback

Piétrain

Vietnamese potbellied pig

Domestic pigs don't have hairy coats to protect them, just a few short bristles. Because of this, pink pigs can get sunburned. Black pigs are protected by their dark skins.

Pig Products

The meat from pigs, which is called pork, is used to make a variety of foods, including sausages, all kinds of hams, and spiced meats.

The last piglet born is often smaller and weaker than the rest. It is called the "runt," and it may not live.

Detective Work

Pigs have a good sense of smell and use their snouts to dig up food. In France, pigs are used to search for truffles, which are difficult to find because they can grow as much as 12 inches (30 centimeters) underground.

Truffles are like mushrooms and are a great delicacy.

A sow usually has about 10 to 12 piglets in a litter. She can have more than two litters every year.

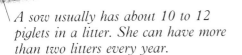

CITRUS FRUITS

The ugli fruit is a cross between a tangerine, an orange, and a grapefruit

The third largest fruit crop in the world, after grapes and bananas, is citrus fruits. This is the family of fruits that oranges, lemons, and satsumas all belong to. Citrus fruits are grown mainly in tropical and subtropical parts of North and South America, in northern and eastern Mediterranean countries, and in Australia.

A tree can produce a thousand oranges each year.

This is a white grapefruit. The flesh inside is a pale yellow.

Lemons, like other citrus fruits, are rich in minerals and vitamins, especially vitamin C.

Tangerines are like small, sweet oranges. They have several small pips.

Limes have thin skins and very sharp juice.

Fighting Frost

Citrus trees need just the right climate to produce good fruit – sun to make them sweet and then cold to make them sharp. But the trees' greatest enemy is frost, which will kill the fruit. So modern plantations often have wind machines and special orchard heaters in case it is frosty.

Orchard heaters protect the trees from frost.

This is a pink grapefruit. Inside, it has a sweet, pink flesh.

The Perfect Package

Citrus fruits are perfectly packaged foods. In the middle is the juicy flesh, a wonderful source of food and drink that is rich in vitamin C. Next comes a spongy layer, called the pith, which cushions the flesh. On the outside is a protective skin.

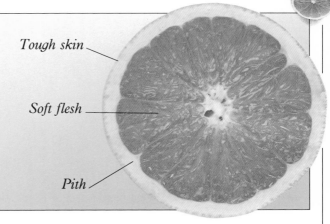

Tough skin

Soft flesh

Pith

Navel oranges are very juicy. Usually they have no seeds.

Clementines, like satsumas and tangerines, have a loose skin that is easy to peel.

Satsumas are like tangerines, except they don't have pips.

You can eat kumquats whole, including the rind.

Mix and Match

By mixing pollen from the blossoms of different citrus trees, new varieties can be developed. The limequat, for example, is a cross between a lime and a kumquat.

Juicy Fruit

These oranges are being sorted and packed and will be sold for eating. Many oranges are pulped to be made into juice. This pulp is frozen for transporting around the world.

Fresh orange juice contains the most goodness, so squeeze it yourself if you can.

GRAPES

There are two different types of grapes: black and white. But black grapes are really dark red, and white grapes are light green. Grapes are delicious to eat fresh and keep their sweetness when dried, too. But most of the grapes grown in the world are used for making wine. Grape plants are called vines. They are planted in rows in vineyards. Vines need a lot of care, and it takes much hard work and skill to produce a bottle of fine wine.

Grapes need rain to swell them and sun to ripen them.

The broad vine leaves shade the grapes from the direct glare of the sun and the battering rain.

Pruning

As vines grow, they are often cut back so they have a few strong branches rather than lots of weaker ones. This work is called pruning and must be done by hand.

Bitter Wine

Red and white wine can also be made into vinegar. It is produced when special vinegar yeasts ferment the wine and turn it sour. Vinegars can be made from other liquids, such as cider or malt, and are used as a flavoring. They can also be mixed with foods to pickle and preserve them.

From Grapes to Wine

First the grapes are crushed and pressed to mix the yeast with the sugar in the grapes. This is called fermentation. Then the juice is filtered and poured into vats. These are left in a cool place for the flavor of the wine to develop.

Grapes   Press

There is a natural yeast on the grape skin that is needed to turn the grape juice into wine.

Black grapes can be made into red or white wine. To make red wine, the skins of black grapes are left in the wine mixture.

Grapes grow in tight bunches.

Dried Goods

Grapes for drying must be very sweet because it is their natural sugar that preserves them. They can be dried on racks, in the sun, or in ovens. Only seedless grapes are used for drying. Sultanas are dried white grapes. Raisins and currants are dried black grapes and are usually smaller.

Currants

Raisins

Sultanas

Grape Harvesters

Most wine grapes today are picked by machine, except in small or sloping vineyards. Special grapes for making rare wines are carefully picked by hand.

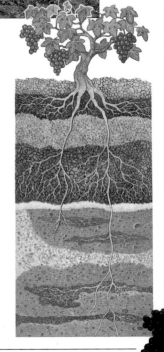

Digging Deep

The type of soil in a vineyard is very important and affects the flavor of the wine. Yet some of the best wines are produced from poor soils. This is because the vine can find regular supplies of food and water by pushing its roots deep into the earth.

Vats

Wine

UNDERGROUND VEGETABLES

Some plants are not grown for their leafy tops but for the parts of them that grow beneath the soil. Carrots, turnips, radishes, and parsnips are the fat roots of plants and are known as root vegetables. Potatoes are the swollen part of the underground stem of the plant. They are called tubers. Onions, too, grow in the dark earth, but it is the bulb of these plants that we eat.

The flowers produce seeds, which can be used to grow new plants, but it is more common to use seed potatoes.

Farming with Nature
Many farmers use chemical fertilizers and pesticides to help them grow crops efficiently, but some farmers use only natural products. They use fertilizers such as manure and encourage pest controllers such as ladybugs, which eat all sorts of flies and larvae. This helps to stop a buildup of chemicals in the environment and is called organic farming.

Potato plants grow well in cool climates, but are killed by frost.

Radish   Carrot   Beet

These are the tubers. Each plant has between 15 and 20 tubers.

Types of Tubers
Tubers come in all shapes and sizes.

Sweet potato

Cassava

Yam

Cocoyam

Harvest Time
In the autumn, the potatoes are fully grown, and special harvesters, pulled by tractors, are used to lift them carefully from the soil.

When the leaves wither in the autumn, the farmer knows the potatoes are ready to harvest.

Tasty Bulbs
Bulbs, such as onions, are not roots. Instead, they are made up of clumps of tightly curled leaves. Leeks, garlic, shallots, and spring onions are all members of the same plant family, and they are all valuable for the flavor that they can give to many other foods.

Garlic

Shallot

Onion

Stem

Some potatoes may be harvested and put to one side. These will then be used to grow new plants and are called seed potatoes.

From these tiny scars, called eyes, new shoots will grow.

Roots

341

HOTHOUSE VEGETABLES

In countries with cool climates, there is not usually enough heat to grow delicate or tropical vegetables. But in special hothouses with glass or clear plastic walls, a farmer can grow almost any crop. This is because the temperature inside a hothouse can be controlled. Hothouses are useful in warm countries, too. They free the farmer from having to follow the seasons, and summer vegetables that do not store well can be grown fresh all year round.

Computer Controlled
Tomatoes are quite fragile plants, so this hothouse, where the temperature and feeding are controlled by a computer, is a good place to grow them. Tomatoes grown in hothouses are usually top quality and are picked by hand.

Nature's "Hothouse"
Some of the world's hottest spots are dry deserts. But if a water supply can be set up, the desert heat can be put to good use. Fertile gardens can be created, such as this fruit orchard in Jordan.

Short Showers
Most of the work in large hothouses today is done by machines. The young sweet-pepper plants shown above are watered by overhead pipes. Hothouse crops must also be sprayed to protect them from the pests and molds that like the damp, warm conditions.

A Carpet of Seedlings

Seeds germinate quickly in the warmth and protection of the hothouse. These are lettuce seedlings. Hothouses can be large, and some are half the size of a football field.

Water Gardens

Some vegetables can be grown without soil. The plants are carefully supported and supplied with water that is automatically mixed with plant foods. This way of growing plants is called hydroponics.

This material does not feed the plants, but helps to support them.

Water full of plant food flows past the roots.

Mixed Vegetables

Here are some other vegetables that would normally need lots of sun to grow, but can be farmed in hothouses in cool climates.

Cucumbers

French beans

Eggplant

Zucchini

Okra

Red and green chilies

Doors must fit neatly to keep out drafts.

These panels may be opened if the hothouse gets too hot.

The hothouse has transparent sides to let in as much light as possible.

This hothouse only uses natural heat, but some have special heaters.

COFFEE, TEA, AND COCOA

Coffee, tea, and cocoa are drunk in most countries of the world. Cocoa is also used to make chocolate. Coffee and cocoa grow in the tropics. They are grown on huge plantations, but a lot is produced on small farms as well. Tea is grown in the tropics and subtropics, on plantations. Altogether, millions of people are employed to produce these crops. Coffee, tea, and cocoa are important to the countries that grow them because they are cash crops. This means they are grown to be traded with other countries rather than for home use.

Picky Pickers

Tea comes from evergreen trees that are pruned into bushes. This makes it easy to pick the leaves. The best quality teas are produced by plucking only the bud and the first two leaves on each shoot. A skilled plucker can pick enough leaves in one day to make 3,500 cups of tea.

Green tea   Black tea

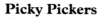

Special Treatment

The leaves are made into green or black tea. For green tea, the leaves are dried, heated, and crushed. For black tea, the leaves are dried, crushed, fermented, and dried again.

A coffee tree produces about 2,000 fruits each year. It takes half this amount to make just one pound (half a kilogram) of roasted coffee.

The fruits are called cherries because of their color and size.

There are two green coffee beans inside each cherry. They only turn brown after they are roasted.

The coffee is harvested twice a year.

The trees are pruned to keep them from growing more than ten feet (three meters) high.

When the fruits first appear, they are green. Gradually, they turn bright red. They are then ready to be picked.

The ripe fruit is picked by hand or shaken from the tree onto a cloth spread out underneath it. On some plantations, machines are used to pick the fruit.

Heavy Load

Cocoa pods are large and heavy. They grow on the trunk of the cacao tree or from its thick branches.

The pod has a thick skin.

Each pod may contain 20 to 50 cocoa beans.

The pod grows up to 15 inches (38 centimeters) long, a few inches longer than a football.

SUGAR

Most people like sugar. Nearly 100 million tons of it are produced each year. Over half of it comes from sugarcane. This is grown on a large scale on plantations in the tropical areas of Brazil, Cuba, India, China, Australia, Mexico, the Philippines, Thailand, and the United States. The rest of our sugar comes from sugar beet, which is a vegetable. This grows in the cooler climates of Europe, the United States, Canada, China, and Japan.

Workers Watch Out!
The sugar plantations can be home to the deadly bushmaster snake. Enormous bird-eating spiders also nestle among the thick foliage.

Sugarcane
Sugarcane is a tropical plant. It grows best in places that get plenty of sunshine and lots of rain.

The canes are cut close to the ground because this is where the most sugar collects.

After the cane has been harvested, healthy stumps will regrow into new plants.

Sugarcane is a very tall grass, like rice and cereals.

The sugar is stored inside the stalk in a firm pulp.

The stalks of a mature plant are about two inches (five centimeters) thick.

Sugar Beet
Sugar beet is a root vegetable. It looks like a giant parsnip. It grows best in places that have warm summers and cool or cold winters.

The leafy tops of the plants are cut off before the sugar beet is lifted. The tops can be used for animal feed.

The sugar is stored in the thick root.

The canes are cut when they are 13 to 17 feet (4 to 5 m) high.

The canes are ripe when they look dry.

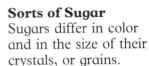

Sugar Production

In a sugar factory, sugar from the canes or beets is taken out, cleaned, and boiled. This leaves a syrup, which turns into brown sugar crystals. These can be refined to make white sugar.

Sorts of Sugar

Sugars differ in color and in the size of their crystals, or grains.

Granulated

Sugarcane is often harvested by machine, but in some places it is still cut by hand.

Turbinado

Maple Syrup

Maple syrup is made from the sap of sugar maple or black maple trees. Holes are made in the trunks in winter, when the trees are dormant. When a thaw follows a freeze, the sap runs and is collected from the wounds in buckets. Maple syrup is produced only in North America.

Dark brown

Muscovado

PEOPLE AND PLACES

In lots of ways, people are the same everywhere: they build homes, they wear clothes, they eat meals, and they have fun. However, men and women in different countries have developed their own languages and customs, which vary widely from place to place.

Did you know, for example, that in Spain, there is a festival in which people throw tomatoes at each other? Did you know that the Chinese language has 50,000 written characters? Did you know that in Denmark, there is a theme park built out of Lego bricks?

Learning about other ways of life is not only interesting, it also helps you understand and get along with people in other countries who may be very different from you.

Inuit boy from northern Canada

African jewelry

The Eiffel Tower in Paris, France

Indonesian boy on a water buffalo

Talking drum
from Nigeria

Antipasto
from Italy

Vegetable market in Guatamala

Buckingham Palace in London, England

Woven bag
from Peru

Indian snake
charmer

THE ARCTIC & ANTARCTICA

The Inuit people live in the icy lands of the Arctic north. For many generations, they have found practical ways of coping with their cold surroundings. But no one has settled for very long on the southern extremes of the planet. Antarctica is the coldest and windiest place in the world. This huge continent is lashed by freezing blizzards and gales, with temperatures as low as -128°F (-89°C).

Icy City
Murmansk is a Russian city 120 miles (200 kilometers) north of the Arctic Circle. No other city is farther north than this. The Sun does not rise between November and the middle of January.

Homes in the North
In the past, the Inuit were self-sufficient and lived a life on the move, hunting and fishing to survive. They built temporary homes, called igloos, from solid snow. Today, most Inuit live in wooden houses and earn a living by working for other people – often in fish-canning factories or for mining companies.

Speedy Snowmobiles
Today, the Inuit people get about on snowmobiles. This is much faster than using dogs to pull a sledge, which is the traditional method of travel.

Warm clothes are vital in the Arctic.

Long-lost Relations
For many years, Inuit families on either side of the Bering Strait were not allowed to meet, for political reasons. Nowadays, the North American Inuit and their relations in northern Asia can visit one another once again.

The Bering Strait is a narrow stretch of sea that separates northern Asia from North America.

The Arctic

Antarctic Science

Scientists from all around the world are the only people who live and work in Antarctica. Some study the effect of the enormous sheet of antarctic ice on the world's weather patterns. Others observe the behavior of living things in the freezing conditions.

This scientist is measuring the density of antarctic snow.

This emperor penguin lives in the Antarctic. No penguins live in the Arctic.

Exploring the Unknown

Antarctica was the last continent to be explored. The first successful expedition in search of the South Pole took place in 1911, led by a Norwegian called Roald Amundsen. A British expedition led by Robert Scott ended in disaster when the team froze to death on the trek home.

What a Mess

Alaska is an important producer of oil. In 1989, an American oil tanker ran aground off Alaska, causing one of the world's worst environmental disasters. Over 1,200 miles (2,000 km) of the Alaskan shoreline were covered with oil.

A windshield protects the riders from the biting wind.

Runners on the front of the snowmobile spread the weight of the vehicle evenly to prevent it from sinking into the soft snow.

CANADA

Canada is an enormous country. A journey from the east to the west coast on the world's longest national highway will take you through almost 4,800 miles (8,000 kilometers) of beautiful landscape. Canada's population of 29 million is small for the country's huge size. Most Canadians live in cities in the warmer southern part of the country, close to the border with the United States.

Totem poles often guarded doorways to village homes.

Some carvings show off the family possessions.

Quebec City

Quebec City is the capital of the province of Quebec. It is Canada's oldest city and the only walled city in North America. Many of its distinctive buildings are in the original French style.

The animals are symbols of the family's ancestors.

Timber!

Half of Canada's 1.6 million square miles (4 million km²) of forest are used for timber. Logging is a very important Canadian industry. On the west coast, large areas of forest are disappearing. Logging companies are being asked to slow down the destruction and to replant more trees.

Totem Pole

The Native Americans of British Columbia in northwestern Canada carved giant totem poles out of trees. Some totem poles celebrated special events or the lives of leaders.

Modern machinery fells giant conifer trees.

Inuit Victory

The Inuit people have lived in the Arctic north of Canada for thousands of years. They have persuaded the Canadian government to give them back control of a large area of their native land. The new territory is called Nunavut and is home to about 17,500 Inuit people.

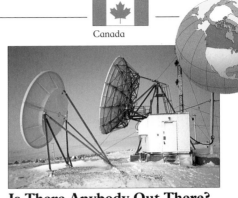

Is There Anybody Out There?

The first long-distance phone call was made in Canada in 1876. A century later, Canada was the first country to set up a satellite network. Satellites provide a vital link to many remote communities.

Two Languages

Canada has two official languages: English and French. The first Europeans to settle in Canada were the French, followed by the British. Today, the majority of Canadians speak English, but French is the official language in Quebec, Canada's largest province.

Winter Sports

Winters are very harsh all over Canada. Ice hockey and skating are national sports. Some families flood their back gardens in winter so that the water freezes to make a temporary ice rink.

A tug pulls the timber down the river. Water transportation is used when there are no major roads through the forest.

The timber is taken to a riverside sawmill, where it is cut into planks or pulped for papermaking.

THE UNITED STATES

The United States is an exciting mixture of different cultures and traditions. Over the last 500 years, about 60 million people from all over the world have made the United States their home, many entering the country through New York City. The families of the new arrivals now help to make up the American population of about 250 million. The United States is a superpower and is probably the richest and most powerful nation in the world.

Leading Light

The Statue of Liberty has been a welcoming sight to many immigrants arriving in the United States over the last hundred years. The statue is a symbol of freedom, and was built to celebrate the 100th anniversary of American independence. It was given to the Americans by the French in 1884.

Lots of Languages

Most Americans can trace their family trees back to other parts of the world. Four out of five Americans speak English, but other European and Asian languages are also widely spoken.

Three-quarters of the American population live in cities. This is San Francisco, in the West Coast state of California.

Jeans are made of tough material, called denim.

Jean Genie

Jeans are an American invention. The first pair was made in 1874 by Levi Strauss. Levi jeans are still sold all around the world. Today, the jeans industry makes millions of dollars for the American economy.

Drive Time

The United States is a huge country, and many Americans are used to driving long distances to visit friends or family. Large cars are not as popular as they used to be. This is because smaller cars use up less gas and are better for the environment. Many cities urge drivers to carpool – or to ride buses instead.

Many American children grow up in suburbs. It is easy to drive from the suburb to the nearest shopping mall or city.

More corn is grown in the United States than in any other country. Much farming is done by machines, so only 3 percent of the labor force works on farms today.

Today's Native Americans

Native Americans live modern lifestyles while still honoring many of their old traditions. Despite widespread decline during the 1880s and later, the population is on the rise.

CENTRAL AMERICA & THE CARIBBEAN

Central America is made up of a narrow strip of eight countries that link the continents of North and South America. Just like the long chain of islands that makes up the Caribbean, Central America is surrounded by clear blue oceans. Today, the beautiful Caribbean Sea and the sunny, sandy beaches attract many tourists.

In the past, the sea brought visitors who were less welcome, like the first settlers from Europe 500 years ago.

Weaving for a Living

Over half the people of Guatemala are Amerindian. Their ancestors go back a thousand years to the days of the ancient Maya. Guatemala now sells Mayan crafts to the rest of the world.

Horizontal weaving loom

The end of the loom is tied to a tree.

Carnival Time

The Caribbean is famous for its carnivals. For two days and nights at the start of the Christian festival of Lent, people dance through the streets wearing fantastic costumes.

It is often possible to tell which village a weaver lives in from the patterns in her cloth.

A strap around the weaver's back holds the loom steady.

356

| Belize | Costa Rica | Cuba | Dominica | Dominican Republic | El Salvador | Grenada | Guatemala |

Pollution City
The 4 million cars in Mexico City cause terrible traffic jams. Pollution from traffic and industry can reach dangerous levels. Mountains around the city stop the bad air from escaping.

The Day of the Dead
Every year in Mexico, a festival is held to remember people who have died. It is called The Day of the Dead and takes place at the end of October. Families have picnics by the graves of their relatives. When night falls, they keep watch over the graveyard by candlelight.

A sugary souvenir of the festival.

A Modern Mix
Many Caribbean people are descendants of enslaved Africans, who were forced to work on sugar plantations. Slavery was stopped in the 19th century, and many Asians then came to work in the Caribbean. Their descendants still live there today.

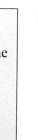

Terrific Temples
Over 1,000 years ago, the ancient Maya peoples built impressive temples, especially in Mexico. Mayan society was very sophisticated long before the Spaniards invaded in the 16th century.

War and Peace
Central America has seen many revolutions and long wars. These children in El Salvador are sheltering from an army helicopter. The civil war in El Salvador ended in 1992, but the peace there is not certain.

 Haiti

 Honduras

 Jamaica

 Mexico

 Nicaragua

 Panama

 St. Kitts-Nevis

 St. Lucia

 St. Vincent and the Grenadines

Trinidad and Tobago

Caribbean People
More than 50 different ethnic groups live on the islands of the Caribbean. Here are just a few.

Asian

Arawak

Afro-Caribbean

European

SOUTH AMERICA

Most South Americans are Roman Catholics. A huge statue of Christ towers over the Brazilian port of Rio de Janeiro. This is a reminder of three centuries of Spanish and Portuguese rule, when the first peoples of South America were almost wiped out. Today, most South Americans still speak Spanish and Portuguese. South America is a growing industrialized area, but many people are still very poor.

Galloping Gauchos

Large areas of northern and central Argentina are covered with a grassy plain, called the pampas. Huge herds of cattle are kept on the pampas and are looked after by cowboys known as gauchos. Most gauchos are descendants of European settlers in South America.

These gauchos from northwest Argentina are wearing traditional Spanish hats and neckerchiefs.

Stiff, flared leather flaps protect the gaucho's legs from high-growing thistles when he is riding his horse in the pampas.

Heavy leather riding crop

Rich and Poor

São Paulo, in Brazil, is South America's biggest industrial city. The city center is very modern and wealthy. Many people from the surrounding areas travel to São Paulo in search of work. Their children often end up living in terrible poverty on the streets.

Colombia

Ecuador

Guyana

Paraguay

Peru

Surinam

Uruguay

Venezuela

Gauchos are very proud of their horses. They work on ranches, called "estancias."

A thick poncho is used as a blanket at night. It was also once used as a shield in a fight.

Machu Picchu and La Paz

South Americans have always been clever at building cities high up in the mountains. The ruins of Machu Picchu are hidden 7,500 feet (2,280 meters) up in the Andes, in Peru. The city was built over 500 years ago, without the help of modern machinery.

La Paz is the capital city of Bolivia. It is the world's highest capital city, 11,916 feet (3,636 meters) above sea level. The city's steep streets are surrounded by snowcapped mountains. The thin mountain air often makes visitors breathless.

The Yanomami

In 1991, an area of the Brazilian rain forest about the size of Portugal was set aside for the Yanomami tribe. This did not stop miners from invading the land, bringing diseases that have wiped out large numbers of the Yanomami population.

South Americans

South America is a huge continent, with a wide mix of different people.

Highland Amerindian

Lowland Amerindian

South American of African descent

South American of European and Indian descent

Algeria	Benin	Burkina	Cameroon	Cape Verde	Central African Republic	Chad	Djibouti

NORTHERN AFRICA

The spectacular pyramids and priceless treasures of ancient Egypt are world famous. Art and culture are still very important to people all over northern Africa. A great range of musical and artistic traditions have been passed down through African families for generations.

The Berber People

Most North Africans are now Muslims. But in Morocco and Algeria, one in five people are still Berbers. Their nomadic way of life has hardly changed for centuries.

Making Music

African rhythms have influenced modern jazz and blues music and are now an important ingredient of African pop music. These drummers are from Ghana in western Africa.

The big bass drums are called brekete drums.

Famine

Africa has the fastest-growing population in the world. It is hard to grow enough food to feed all these people. Even in a normal year, up to 150 million Africans suffer from a lack of food. When the rains don't come, crops fail and disaster strikes. These children in the Sahel region of northern Africa are in danger of starving.

Egypt
 Ethiopia
Gambia
Ghana
Guinea
  Guinea-Bissau
Ivory Coast
Liberia

City Growth
Most Africans live in the countryside, but a steady flow of young men are moving from the villages to look for work in the big cities. Lagos in Nigeria is Africa's fastest-growing modern city.

Chocolate Center
Well over half the world's cocoa is grown in western Africa. More cocoa is grown in the Ivory Coast than anywhere else. About a fifth of Ivory Coast farmland is used to grow cocoa crops.

The drummers' clothes and music come from Dagbon, in northern Ghana.

The smaller drums are called talking drums. The tighter the strings are squeezed, the higher the note.

African Crafts
Wood carving, weaving, and jewelry making are traditional African crafts.

Nigerian wood carving

Narrow-strip woven trousers from Ghana

Bronze bracelet from Mali

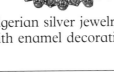
Algerian silver jewelry, with enamel decoration

Brass bracelet and earrings from Sudan

 Libya
 Mali
 Mauritania
 Morocco
 Niger
 Nigeria
 Senegal
Sierra Leone
Somalia
 Sudan
 Togo
Tunisia

(Flag not shown.) Eritrea

361

SOUTHERN AFRICA

Angola

A rich variety of peoples make up the African countries in this region. Huge grassland savannahs support an amazing variety of wildlife. Game parks were first set up in Kenya to protect animals from hunters. Today, these parks are big tourist attractions. Across the whole region, traditional lifestyles, like those of the Kalahari Bushmen, have slowly been replaced with new ways. There have been many personal and political changes for the peoples of southern Africa.

A New Start
The San, or Bushmen, live in the Kalahari desert. For about 20,000 years, they hunted animals and gathered plants to survive. Very few of them still hunt, and most now find work on local farms or in nearby towns.

Today, most Bushmen settle in one place. They have given up a life on the move.

On the Move
Only a few small groups of Bushmen still hunt in the Kalahari desert. Each group has a territory of up to 400 square miles (1,000 km²). The men hunt while the women find tasty fruit, nuts, and roots.

Much of the Bushmen's land is now used for farms, cattle ranches, and nature reserves.

tswana

Burundi

Comoros

Congo

Equatorial Guinea

Gabon

Kenya

Lesotho

Madagascar

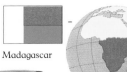

Market Day

African women often earn money by selling surplus vegetables from the family farm at the market. Village markets are lively and noisy occasions, where friends meet to exchange the latest news.

Malawi

Mauritius

Mozambique

Namibia

Rwanda

São Tomé & Príncipe

South Africa

Going to School

Many African children have a long journey to school. A bus takes these Zimbabwean children part of the way, and they have to walk the last mile or so.

Black and White

Nelson Mandela, leader of the African National Congress, was jailed for 28 years while fighting against apartheid for the rights of South Africa's nonwhites. The last apartheid laws were repealed in 1992. Released from prison in 1990, Mandela was elected South Africa's president in 1994.

Swaziland

Tanzania

Uganda

Zaire

Rise of the Zulus

Two hundred years ago, the Zulus were a small clan of a few hundred people, but they fought wars with similar clans to become one big Zulu nation. Today, the Zulus are an important political force in South Africa.

White Tribe

This Afrikaner family lives on a farm in South Africa, near the border with Zimbabwe. They are descendants of the first Dutch South African settlers, who started arriving from Europe in 1652. Like their ancestors, the Afrikaners are still dominant in many aspects of South African life.

Zambia

Zimbabwe

THE MIDDLE EAST

The discovery of oil has made parts of the Middle East very rich and powerful. Huge oil refineries are a common sight in the region. But the oil will not last forever, and other industries are being developed to keep the money coming in when the oil has run out. The Middle East has always been a very productive area. Its age-old civilizations have exported farming techniques, beautiful crafts, and three major religions all around the world.

Desert Survivors
This family lives on the edge of the desert in Oman. They belong to a group of traveling Arabs, called the bedouin. For 4,000 years, nomadic bedouin have survived in the hottest parts of the Middle East, leading camels and goats in search of water and pasture.

Bedouin women dress modestly in traditional masks and black overdresses.

Bedouin girls wear colorful clothes and silver and gold jewelry.

Turkish Bazaar
In the Middle East, many people shop in markets and bazaars. This shopkeeper is hoping to sell some of his brassware at the Great Covered Bazaar in Istanbul, Turkey.

It's a Camel's Life
Most bedouin families own a pickup truck. These do much of the work once done by camels. Camels are prized as symbols of wealth. A camel may be sold to a camel racer for several thousand dollars.

Living on a Kibbutz

In Israel, big farms, called kibbutzim, were set up so many Jewish families could live and work on the same farm and share the produce.

The Dome of the Rock

City of Faith

Jerusalem is a special city for Jews, Christians, and Muslims. Jews come to pray at the Wailing Wall, on the site of the Old Temple. Christians believe Christ came back to life after his death in Jerusalem. The Dome of the Rock marks the spot where Muslims believe Muhammad rose to heaven.

The Wailing Wall, where Jews go to pray.

A decorative silver dagger is a sign of power and wealth.

White clothes reflect bright sunlight and help people keep cool.

Bedouin men and boys wear head shawls to protect their heads from the sun. In June, the Omani desert can reach 113°F (45°C) in the shade.

Trip of a Lifetime

At least once in a lifetime, every Muslim must try to make a special journey, or Hajj, to Mecca in Saudi Arabia. During the Hajj, visitors crowd into the Great Mosque and walk seven times around the shrine, called the Ka'ba.

Afghanistan

Bahrain

Iran

Iraq

Israel

Jordan

Kuwait

Lebanon

Oman

Qatar

Saudi Arabia

Syria

Yemen

United Arab Emirates

Turkey

SCANDINAVIA

Scandinavia is the name given to the countries of northern Europe. The far north of this region lies inside the Arctic Circle, so the winters are very harsh. The people of northern Scandinavia enjoy winter sports in the cold climate. Cross-country skiing was invented in Norway and is still a quick way of getting about during the snowy winter months.

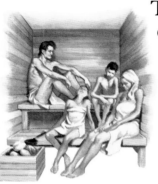

City Slickers
About one-quarter of the Danish population live in Copenhagen, the capital of Denmark. Most Scandinavians choose to live in cities in the warmer south of the region.

Sweat It Out
Saunas are wooden rooms steaming with heat from a stove. Invented in Finland over 1,000 years ago, they are found in many homes in Scandinavia. After a sauna, people often enjoy a cold shower or a roll in the snow.

Escape to the Country
Many Scandinavians enjoy a high standard of living. Some can afford to buy or rent a second home in the mountains, in the forest, or by the sea. They escape to these "holiday huts" on weekends and during the summer months.

Dark Days
It is the middle of winter, and the people in this northern Scandinavian town are eating midday lunch in the dark! In winter, the tilt of the Earth causes countries in the far north to get less of the Sun's light. Days get shorter, until the Sun doesn't rise at all.

Land of the Midnight Sun
Imagine waking at midnight to find the Sun shining outside! At the height of summer, the tilt of the Earth causes northern countries to get more of the Sun's light. In northern Scandinavia, it doesn't get dark at all in the middle of summer.

It is still light at midnight, because the Sun hasn't set.

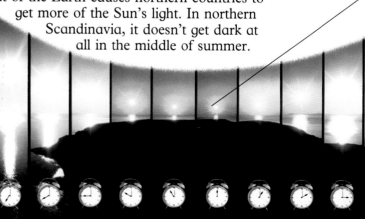

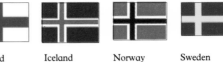

Denmark Finland Iceland Norway Sweden

Out in the Cold

The Saami, or Lapp people, are traditionally a nomadic group, traveling throughout northern Scandinavia and Russia herding their reindeer. It is getting harder and harder to make a living in this way, so many young Saami are now becoming fishermen and farmers.

Legoland

Denmark is the home of Lego, one of the world's most popular children's building toys. A theme park has been built entirely from Lego bricks near Billund. It is now a big tourist attraction.

Lego giraffe

Earthly Energy

These Icelanders are swimming in waters warmed by energy deep under the Earth. This geothermal energy is also used to generate electricity and heat houses.

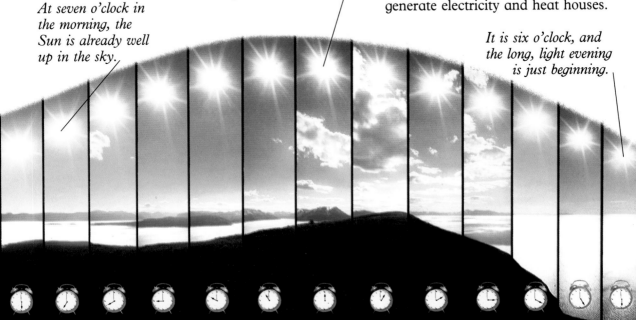

At seven o'clock in the morning, the Sun is already well up in the sky.

At midday, the hot Sun is at its highest position in the sky.

It is six o'clock, and the long, light evening is just beginning.

THE BRITISH ISLES

The Republic of Ireland is an independent country, with its own government and traditions. England, Scotland, Wales, and Northern Ireland make up the United Kingdom. Many aspects of British life, such as its historical buildings and customs, have been preserved. But the United Kingdom is also good at adapting to change. A rich variety of ethnic groups now live in the United Kingdom's multicultural society.

Beside the Seaside
Britain is surrounded by the sea. British seaside holidays first became popular in the 19th century. Today, many British people go on vacation abroad, where it is less likely to rain!

The Oldest Democracy
The Houses of Parliament in London have been the home of British democracy since 1512. The political party with the most representatives, called MPs, forms the government. MPs are voted in at national general elections.

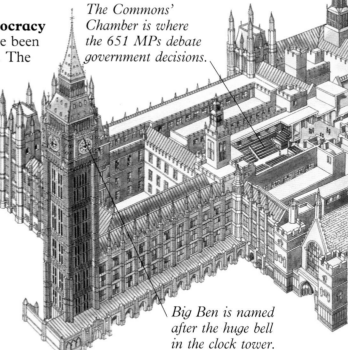

The Commons' Chamber is where the 651 MPs debate government decisions.

Big Ben is named after the huge bell in the clock tower.

Village Cricket
Sunday afternoons in summer would not be the same in many English villages without the familiar sight of a cricket match on the village green. Cricket is an English invention and is usually followed by another well-known English tradition, afternoon tea!

Guy Fawkes Night
A plot to blow up the Houses of Parliament almost succeeded in 1605, but the ringleader, Guy Fawkes, was caught and executed. Guy Fawkes Night is still celebrated all over England on November 5, when home-made straw dummies – called guys – are burned on big bonfires at fireworks parties.

Arty Edinburgh

For three weeks every August, a big arts festival takes place in Edinburgh, the historic capital of Scotland. The festival gives amateur and part-time entertainers the chance to perform and also attracts some of the world's best musicians, theater companies, and artists.

Foot-tapping Folk

Traditional folk music is extremely popular in Ireland. The accordion is a typical folk instrument, often played to accompany singers. Irish bands have been very successful on the world pop scene. Many songs were first written in Gaelic, the original language of Ireland.

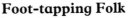

Possible new laws are debated in the Lords' Chamber.

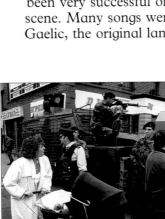

A Divided Community

Northern Ireland is part of the United Kingdom, but is divided along religious lines. The Protestant majority want to stay under British rule, while the Catholic minority want to join the Republic of Ireland. Police and the British Army try to keep the peace in this violent conflict, while ordinary people try to carry on with their lives.

National Pride

The Welsh are proud of their culture and enjoy celebrating national festivals. St. David is the patron saint of Wales, so on St. David's Day, children go to school wearing traditional Welsh costume.

The leek is the national emblem of Wales.

FRANCE & THE LOW COUNTRIES

The Netherlands, Belgium, and Luxembourg are sometime called the Low Countries because much of the land is flat and low-lying. France and the Low Countries are powerful farming and trading nations, thanks to their good soil and large natural harbors. The people of the region are fond of good food. French chefs are world famous for their fine cooking skills.

European Dream
The European Union flag represents its 15 members: Austria, Belgium, Denmark, Finland, France, Germany, Greece, Italy, Luxembourg, the Netherlands, Portugal, the Republic of Ireland, Spain, Sweden, and the United Kingdom.

This stylish outfit is by Christian Lacroix, a top French fashion designer.

Style Capital
Paris, the capital of France, is at the heart of the world fashion industry. Every season, famous designers display new creations at big shows.

Chocolate Secrets
Many Belgian chocolate makers keep their recipes secret so no one can copy them.

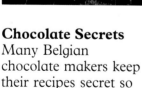

Brandy

Coffee

Mint

Nougat

Mixed nuts

City Canals
Amsterdam is the capital of the Netherlands. It is built around a network of canals that were once used for trade and transportation. By tradition, the Dutch find their way around the city on bicycles.

At War with the Sea
Nearly half of the Netherlands was once under seawater. Over the centuries, the Dutch have reclaimed this land from the sea, draining it by using a clever system of canals and seawalls, or dikes. Much of the reclaimed land is now used for intensive crop and dairy farming.

| Belgium | France | Luxembourg | Monaco | Netherlands |

Land of Wine

France produces some of the world's best wines. Each region has its own wines, with their own special flavors. In the Rhône valley in southeast France, grapes ripen in the summer sun. They are then harvested and fermented in vats.

Small but Rich

Luxembourg is a tiny country that nestles between France, Belgium, and Germany. Its small population has a high standard of living because of the country's success as a financial center.

Fabulous Food

French cooking is world famous, and French people take care to buy food as fresh as possible. They are careful shoppers, closely inspecting food and even sampling it before they buy.

Attractive vegetable displays are arranged to catch the shopper's eye.

Many French shops sell only one kind of produce, but offer a wide range of choices.

Crusty French bread is bought every morning because it does not keep fresh for long.

371

SPAIN & PORTUGAL

Both Spain and Portugal have warm, sunny climates and long coastlines. The blue skies and dazzling white beaches of Spain's Mediterranean coast attract millions of tourists every summer. The sea is also an important source of food and employment. Fishing is a major industry in Portugal and on Spain's Atlantic coast.

Fiesta!
Spain has more festivals than any other European country. Some are religious. Others are just for fun. One of the messiest festivals takes place in Bunyols, in Valencia. The villagers pelt each other with tomatoes to celebrate the time when a truck spilled its load of overripe tomatoes all over the village square.

Easy Riders
Scooter riding is a craze for young people in many parts of Spain, especially the regions of Valencia and Catalonia. Bikers meet in the evenings to compare bikes and arrange races.

Young Spaniards can ride scooters from the age of 14.

This scooter can reach a top speed of 60 miles (95 km) an hour.

Tide of Tourists
Mediterranean countries are very popular with tourists. In 1990, 170 million tourists flocked to the coasts of southern Europe. Every year, the Spanish population doubles when tourists arrive to enjoy the hot summer weather and sandy beaches.

Andorra   Portugal   Spain

Take Your Pick

These Portuguese farmers grow oranges, lemons, and olives and are especially proud of this year's crop of oranges. Orchards of sweet, juicy oranges are a common sight in the hot countryside of southwest Portugal.

City of Seven Hills

Lisbon, the capital of Portugal, is the smallest capital in Europe. It is built on seven steep hills. Old electric trams carry the townspeople up and down the sloping streets.

Eating around Spain

Here are some tasty dishes from different regions of Spain.

Gazpacho – cold tomato soup from Andalusia

Fire Dance

Flamenco dancing was developed many hundreds of years ago by the gypsies of Andalusia. It is still popular all over Spain. Many Spanish children learn the dance from a young age and enjoy wearing the colorful flamenco costumes.

Cocido – meat stew from Castile and Estremadura

Zarzuela de Pescado – seafood stew from Catalonia

Paella – chicken, seafood, and rice from Valencia

First Communion

Most Spanish people are Roman Catholics. From the age of seven or eight, children are taught the main beliefs of the Catholic church, ready to take their first communion at the age of ten. The ceremony is a religious and social occasion and is taken seriously by all the family.

Doughnuts – a popular sweet all over Spain

GREECE & ITALY

Greece and Italy lie in southern Europe, where the warm summers and beautiful Mediterranean coastline draw millions of tourists every year. Many magnificent buildings, dating back to the days of the ancient Greeks and Romans, are also popular attractions. A lot of these sites are in urgent need of repair. Both Greece and Italy are struggling to keep their ancient monuments in good condition.

Priests on Parade
Greek Orthodox priests are allowed to get married, unlike Roman Catholic priests. Priests play an important part in both Greek and Italian society and are given great respect.

Café Talk
In some parts of Greece and Italy, far away from the busy towns and cities, life carries on at a very relaxed pace. Many families rest in the strong heat of the day. In the cooler evenings, men often gather in cafés for a drink and a chat.

Chemicals from car exhausts are eating away at the ancient stone.

Parts of the Parthenon have been so badly damaged, they need to be restored or replaced.

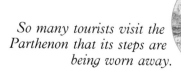

So many tourists visit the Parthenon that its steps are being worn away.

Temple under Threat
The Parthenon in Athens is an ancient Greek temple, built to honor Athena, the goddess of wisdom. It is now under threat from air pollution and the tramping of millions of visitors' feet.

Cyprus

Greece

Italy

Malta

San Marino

Vatican City

The Pope's City
The Pope is head of the Roman Catholic Church. He lives in a tiny country called the Vatican City, which lies within Rome, the capital of Italy. This state has its own laws and police force – and its own postal service, too!

Car Italia
The Fiat car works in Turin, northern Italy, is one of the largest car factories in the world. Over 140,000 people work there to make more than one and a half million cars every year.

Pasta Preparations
More types of pasta are made in Italy than anywhere else in the world. Flour, eggs, vegetable oil, and salt are mixed to make dough, which is then rolled out and cooked. Pasta is often eaten with a tasty sauce.

Spaghetti Bolognese

Big arches reduce the weight of this typical Venetian building.

Brick foundations are laid on wooden posts, which are driven deep into the mud.

Grand Passions
Opera and soccer are two of Italy's top obsessions. The soccer World Cup took place in Italy in 1990. Pavarotti, the world-famous Italian opera singer, sang at a gala concert to celebrate this important tournament.

Sinking City
Venice was built over 300 years ago on the mudbanks of a lagoon on Italy's northeast coast. The floating city is slowly sinking. Many of its buildings have been damaged by pollution and constant contact with water and need to be carefully restored.

GERMANY, AUSTRIA, & SWITZERLAND

Germany, Austria, and Switzerland lie at the heart of Europe. Central Europeans share a love of traditional food, drink, and festivals. They also share a language: German is the most widely spoken language of the region. Germany is a powerful country. After years of separation for political reasons, East Germany rejoined West Germany in 1990 to become one big country. The wall dividing East from West Berlin was torn down amidst great celebrations.

Fantasy Castle

The castle of Neuschwanstein nestles among the Bavarian Alps in southern Germany. It was built about a hundred years ago by a rich Bavarian king with a wild imagination.

King Ludwig II died in 1886, before this spectacular castle was finished.

Tall turrets give the castle a fairy-tale appearance.

Starting School

German children start elementary school at the age of six, although most have been to a nursery, or kindergarten, for three years before this. On their first day, children are sent to school with a large cone stuffed with everything they will want, including candy, pens, and books.

The castle attracts tourists from all around the world.

Austria

Germany

Liechtenstein

Switzerland

Coming Down the Mountain

The return of cows from the Alps is celebrated in August. It is called the Einabtrieb. The cows are usually decorated in bright colors. But if somebody has died, the decorations are black or dark blue.

Snowy Holiday

Every year, thousands of skiers visit resorts in Austria and Switzerland. Some people worry that mountain tourism spoils the environment. But it brings lots of money for both countries and is a big employer of local people.

Very Viennese

One in five Austrians lives in Vienna, the capital of Austria. The city was once at the center of the great Austro-Hungarian empire. It is still an important city as home to many United Nations organizations.

The highest point of the wheel is nearly 215 feet (65 meters) above the ground.

Each gondola takes up to 12 passengers.

The Big Wheel

The Riesenrad in Vienna is one of the largest Ferris wheels in the world. The wheel is 200 feet (61 meters) across, and it turns around at the slow speed of 30 inches (75 cm) a second.

Three Languages

Switzerland is surrounded by Germany, Austria, Liechtenstein, France, and Italy. Three main languages are spoken in Switzerland – French, German, and Italian. Most Swiss people speak German, but many speak at least two of these languages.

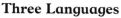

EASTERN EUROPE

Eastern Europe has been invaded many times and is now a meeting place for many different cultures and religions. The breakdown of Communism during the 1980s and 1990s has brought new freedom to the region, but has also allowed back old tensions between many of the ethnic groups.

War-torn
A wide mix of people used to live together peacefully in former Yugoslavia. At the end of the 1980s, war broke out between the different ethnic groups. A once beautiful European country has now been torn apart by the conflict.

Czech Cartoons
Children's films from Slovakia and the Czech Republic are world famous. This Czech film shows a typical mix of live action and puppet animation.

Royal Romanies
The Romany people are also called gypsies. This group is the "royal family" of the gypsy king of Romania. There may be well over 250,000 gypsies in Romania.

The Romanies are one of the biggest minority groups in Romania. This man is the king of the Romanian gypsies.

Gypsies have always been independent, refusing to live in cities and change their traveling way of life.

Albania

Bosnia and
Herzegovina

Bulgaria

Croatia

Czech Republic

Hungary

Macedonia

Poland

Romania

Slovakia

Slovenia

Yugoslavia

Dracula's Castle

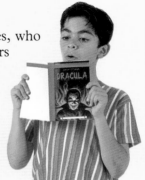

A Romanian prince, Vlad Tepes, who lived in this castle over 500 years ago, inspired the story *Dracula*. This story about a scary, bloodsucking vampire is not true. But the real prince who lived here in the Transylvanian Alps was indeed a cruel man who killed all his enemies.

Hungarian Farming

Farms like this one in Hungary cover three-quarters of the country's land. Modern farming methods have not yet reached much of eastern Europe. Crops have been harvested in the same way for many generations.

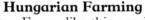

Many gypsies speak Romany, which is similar to some northern Indian languages.

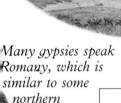

City of Learning

Cracow was once the capital of Poland, and its streets are lined with many beautiful buildings. Cracow is Poland's oldest university city, founded over 600 years ago.

Pollution Problem

Many countries in eastern Europe suffer from bad pollution. Not enough money is spent on modernizing old factories. Foreign industries have set up new factories, but some of these have made the pollution problem even worse.

Favorite Foods

These are some eastern European dishes.

Cherry soup

Sauerkraut

Goulash

Apple strudel

NORTHERN EURASIA

Northern Eurasia stretches over a vast area, from the Ukraine in the west to the frozen wastes of Siberia. For centuries, northern Eurasia has been controlled by powerful rulers and governments. But recently, the communist Soviet Union split into 15 independent states. Mongolia had close links with the former Soviet Union, but was never part of it.

Daring Dance
This energetic dance celebrates the courage of cossack soldiers from the Ukraine.

Rocking Russia
These rock fans are enjoying a concert in Moscow, the capital of Russia. Big changes happened for the Russians in the late 1980s. Strict communist rules were relaxed, and the people began to enjoy new freedoms.

Going to Church
Orthodox Christianity has been the traditional Russian religion for over 1,000 years. The cathedral of St. Basil is one of the most spectacular Russian Orthodox churches.

The cathedral lies in Red Square, right in the heart of Moscow.

Keeping the Faith
The Communist Party disapproved of all religions. The Jews suffered badly – their synagogues and schools were closed down. Jewish traditions survived, but they are now practiced mainly by older people.

St. Basil's was built for Tsar Ivan the Terrible in the 1550s to mark his military successes.

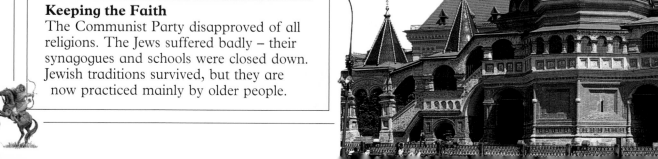

High Fliers

The circus is a popular form of entertainment all over Russia. One of the most famous circuses in the world is based in Moscow. The New Moscow Circus boasts a high standard of acrobatics. Its grand Big Top was opened in 1971 and seats an audience of up to 3,400 people.

Northern Eurasians

The largest ethnic group of northern Eurasia is made up of Slavs. Next biggest is the Turkic-speaking population. Smaller groups include Mongols and Inuit.

Eastern Promise

The people of Central Asia look very different from the Russians who live farther north. They speak languages similar to Turkish and many are Muslims, like their neighbors in the Middle East. These women are selling bread at a market stall in Uzbekistan.

Ukrainian Slav

Turkic-speaking Uzbek

Yesterday's Hero

Ancestors of these Mongol horsemen fought in the army of the fierce Mongol emperor Ghengis Khan 1,000 years ago. Today, Ghengis Khan is once again a Mongolian national hero.

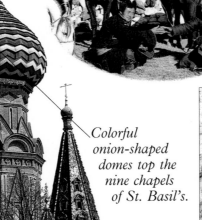

Colorful onion-shaped domes top the nine chapels of St. Basil's.

Mongol

Ripe Old Age

This Azerbaijani couple married 102 years ago and have 190 descendants! A lifetime of fresh mountain air and a healthy diet of honey, herbs, and dairy food has seen them well into old age.

Russian Inuit

CHINA, HONG KONG & TAIWAN

There are more than 200 million bikes in China, so cycling in the rush hour can be very slow! There are more people in China than in any other country – over a billion people live and work in this amazingly varied land. Fifty-seven different ethnic groups make up the population, but by far the largest group is the Han, traditionally a peasant farming people.

Communism in China

The Communist government in China plays an active role in people's everyday lives. This official government poster encourages couples to have only one child.

Food and Farming

Two-thirds of the Chinese people are farmers, growing crops on every spare patch of suitable land. Up to three crops of rice may be grown on the same paddy field each rice-growing season.

Rice is the basic food for China's huge population.

Bringing up Baby

The "one child per family" rule was introduced in 1979 to try to control the fast population growth. It was needed to keep the Chinese population under 1.2 billion by the year 2000.

Words and Pictures

The Chinese written language is based on pictures, which describe objects, actions, and ideas. There are about 50,000 pictures, or characters. A simplified list of about 2,000 is commonly used today.

女人
Woman

男人
Man

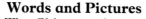

Lettering brush

China Taiwan

Folk Art in Taiwan

Taiwan has some colorful folk traditions. A sticky mixture of rice and flour is dyed and molded to make these decorative figures. In the past, the flour and rice recipe was used to make children's snacks. Today, the figures are kept, not eaten.

Many Chinese people would like to be allowed to have bigger families. It is hard for them not to spoil their "only" children.

Hong Kong

Nearly 6 million people live in the 418 square miles (1,045 km^2) of Hong Kong, making it the world's most crowded place. Hong Kong is a busy trading center and the world's biggest exporter of watches, clocks, and toys.

Get Well Soon

This street trader is selling herbal medicines. Ancient, natural ways of treating illnesses are very common in China.

Angelica root

Most Chinese mothers work full time. Their children are looked after in day care centers at the workplace, until they are old enough to go to school.

Buddhist Tibet

Most people living in Tibet are Buddhists. They believe their past leaders are alive now in the body of their present leader, the Dalai Lama. In 1959, the Dalai Lama fled his home, the Potala Palace, to escape the invading Chinese army. He now lives in exile in India.

JAPAN & KOREA

Japan is famous for its electronic gadgets and machines. In the last 35 years, Japan has become one of the world's richest countries, and many Japanese people enjoy a high standard of living. Japan's neighbor, Korea, is split into two separate countries. North Korea has cut off its links with South Korea and is one of the world's most secretive societies. South Korea is a successful industrial country, like Japan, and trades with the rest of the world.

Fish Food
The traditional Japanese diet of fish is very healthy. Many Japanese people live to a very old age – in 25 years' time, about a quarter of the population will be elderly.

Children on Parade
The Seven-Five-Three Festival in Japan is named after the ages of the children who take part. Every year on November 15, girls and boys are dressed up in traditional costume and taken to a Shinto shrine. It is a very social occasion.

Dinner Time
The evening meal is an important time for busy Japanese families to get together and relax. Parents often get home late from work, and children have lots of homework to do in the evenings.

Rice is cooked until it is sticky, so it is easy to eat with pointed chopsticks.

An electric rice cooker sits close to the table.

Mind Your Head
These pupils are not hiding from their teacher! They are practicing what to do if there is an earthquake. Earthquakes do not happen often in Japan, but they could happen any time. In the past, they have caused terrible damage.

Japanese tea is drunk without milk and is called green tea.

書店

Colorful Koreans

This woman is wearing the national costume of South Korea. It is made of colorful material. South Korea sells huge amounts of material to the rest of the world. About 720 million square yards (600 million m²) of cotton were produced in South Korea in 1990.

The food is placed on the table before the family sits down to eat.

The family sits on cushions around a low table to eat their meal.

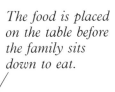

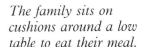

Selling Ships

Most of the ships in Asia are made at the Hyundai shipyard in South Korea, the biggest shipyard in the world. In 1990, 35 big ships were built at the yard – altogether, they weighed a massive 1.8 million tons.

Watch This Space

The big cities of Japan can get extremely crowded. Space is often very hard to come by, but the smallest places are always put to good use. This shopkeeper has hundreds of personal stereos for sale in a tiny kiosk in Tokyo.

Chopstick rest

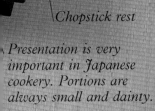

Presentation is very important in Japanese cookery. Portions are always small and dainty.

Climb Every Mountain

Japanese people work hard, but they also enjoy their hobbies. Japan has some stunning mountain ranges, and many Japanese people keep fit with mountain climbing. These hikers are up Mount Furano-Dake, on the island of Hokkaido in northern Japan.

書店

THE INDIAN SUBCONTINENT

The Indian subcontinent includes India, Sri Lanka, Pakistan, Bangladesh, Nepal, Bhutan, and the Maldives. Most people in this part of the world are Hindus, Muslims, or Buddhists. In India, 80 percent of the huge population is Hindu. Every year, millions of Hindus make pilgrimages to the holy city of Varanasi in northern India to wash away their sins in the Ganges.

Country Life

Eight out of ten Indians live in small country villages. Cows are kept in each one, and the human population is only three times bigger than the cow population! Hindus believe that cows are holy, so they don't eat them. According to Hindu teaching, 330 million Hindu gods and goddesses live in the body of a cow.

The bride is nine years old. After the wedding, she will return to her family until she is old enough to live with her new husband.

The beautiful wedding sari is made of expensive silk.

The people in this village make combs for a living.

Wedding Day

Most Hindu parents choose the person their son or daughter will marry. It is important that the partner they pick belongs to the right class, or caste, of Indian society. Children are often engaged and married when they are quite young. Their weddings are colorful and happy family occasions.

Hindu Theater

The Dusserah festival celebrates the end of the rainy monsoon. Groups of actors perform episodes from the story of Ram, a Hindu god in human form. The main parts are played by the top class of Indian society, called the Brahman caste.

Bangladesh
 Bhutan
India
Maldives
 Nepal
 Pakistan
Sri Lanka

The bride's family gives expensive presents to the groom's family. This is called the dowry.

The groom is 11 years old. He is leading his bride to the marriage ceremony.

Floods in Bangladesh

Bangladesh is a very low, flat country, crisscrossed with rivers. After heavy rain, flooding can cause terrible damage. This farmer is helping to push a tricycle taxi, called a rickshaw, through a flooded village in northeastern Bangladesh.

Women at the Top

Most people in Pakistan are Muslims. In 1988, Benazir Bhutto was voted the first prime minister of a Muslim country. In 1993, she was elected for the second time.

Climb Every Mountain

The Sherpa people of Nepal live high up in the Himalayas. Many Sherpas earn money by guiding tourists, hikers, and mountaineers up the steep and rocky paths.

Red-hot Spice

Curry is a very popular Indian dish and is enjoyed all around the world. Spices are important ingredients in a curry.

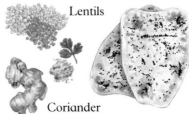

Lentils

Coriander

Ginger

Nan bread

Spicy potatoes

Lamb curry

SOUTHEAST ASIA

Brunei

Burma

It is hard to get away from water in most parts of Southeast Asia! The ten countries that make up the region are linked by waterways and surrounded by oceans. The hot, wet environment is ideal for farming rice, the staple crop of the region. It also supports large areas of tropical rain forests. Sadly, heavy logging has forced many forest people to move away to towns, where the way of life is very different.

People Power
The Penan people live in the rain forests of Sarawak. They have tried to stop the destruction of their forest home by blocking roads into the rain forest. Sarawak has the world's highest rate of logging.

The Secret of Success
The small island state of Singapore is a powerful trading nation. Singapore became successful because of its position. It was built up as a key trading post between the Far East and the West, and is now one of Asia's richest countries.

Living with Water
Many houses in Malaysia and Indonesia are built on stilts, which helps protect them from floods during torrential rainstorms.

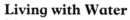

The houses are called "long" houses. Behind each door is a large room, or "bilik," where the family lives.

The houses are made from both natural and modern materials. Some roofs are made from palms, others from corrugated iron.

ambodia
 Indonesia
 Laos
 Malaysia
 Philippines
 Singapore
 Thailand
 Vietnam

Floating Market

Much of Bangkok in Thailand is below sea level, and the city has a busy network of canals. Floating markets are colorful occasions – local traders bring their goods to market by boat, and shoppers paddle up to take a look at what is for sale.

Bali Dancers

The Indonesian island of Bali is famous for its exciting dances. Historical tales of local princes and heroes are acted out in a masked dance, called the Wayang Topeng. The dance is very entertaining, with lots of clowning about as well as serious storytelling.

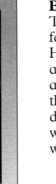

A Topeng dancer acts out the character of the mask.

Every movement is carefully controlled.

Doing Time

The vast majority of Thais are Buddhists. Every male Buddhist is expected to become a monk for a while to study religious teachings and prepare for adult life.

Pirates on the South Seas

Pirates are still terrorizing the seas of Southeast Asia, capturing boats and stealing their cargo. In March 1991, a big oil tanker disappeared off the Philippines. This was almost certainly the work of pirates.

Boats are the main form of transportation. Rivers provide links between small communities.

Notched tree trunks or ladders are used to get down to the water level.

AUSTRALASIA & OCEANIA

Most Australians live in big cities like Sydney, the capital of New South Wales. Both Australia and New Zealand are big economic powers, trading mainly with Asia and the United States. By contrast, many Pacific Islanders of Oceania live in isolated communities that have little contact with the outside world.

Emergency!
The Royal Flying Doctor Service is a lifesaver for people living in remote parts of Australia. Few patients are more than a two-hour flight away from a hospital.

Australian Megamix
British settlers first came to Australia about 200 years ago, but in the last 50 years there have been many new arrivals from other parts of Europe and Asia.

A large number of Australians have British ancestors.

New European arrivals include many Greeks.

Many Italians moved to Australia after the Second World War.

Aboriginal Art
Aboriginal peoples were the first to settle in Australia, over 40,000 years ago. They have strong artistic traditions, and Aboriginal art is now sold around the world, making much needed money for their communities.

The majority of Australians share a love of the outdoors and live by the sea.

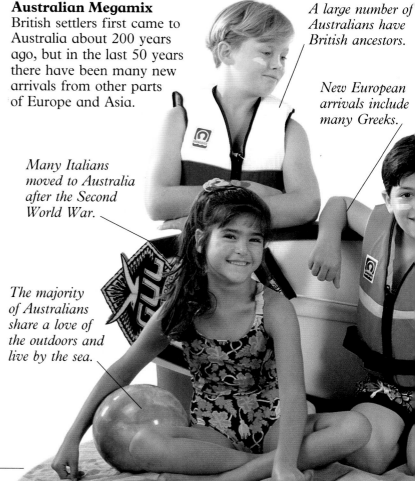

Nauru

New Zealand

Papua New Guinea

Solomon Islands Tonga

Tuvalu Vanuatu Western Samoa

Teaching in Tonga

Tonga is one of the independent nations among the Pacific Islands of Oceania. English and Tongan are the two official languages for the small population of 95,000. In Tonga, children have to go to school between the ages of 6 and 14.

First People

These people were some of the first settlers in Australasia and Oceania.

Pacific Islander

This boy has Vietnamese parents. English is his second language, just as it is for one in eight other Australians.

Stone Money

Some Pacific island tribal traditions have not changed for thousands of years. The people of Yap Island still use stones like this one as money when they make important property deals.

Papua New Guinea tribesman

This girl's parents are Chinese. Many Australians are of Asian descent.

Maori woman

Rugby Dance

The Maoris settled in New Zealand about 1,000 years ago. The New Zealand rugby team has borrowed some Maori traditions. They use an old Maori war chant, called a haka, to get them in the mood for matches.

Surfing is popular along the coast of Australia.

Aboriginal tribesman

CHAPTER 4

SCIENCE AND TECHNOLOGY

Through the centuries, people's lives have changed enormously. Today, the way you live is completely dominated by technology, from the electric toaster that produces your breakfast to the bus that takes you to school, the computers you use for learning and playing games, and the television and video you watch in the evening.

Science and technology have helped us in lots of ways. Because of them, we can grow and distribute food cheaply, manufacture clothing and household goods in huge quantities, treat illnesses of all kinds, and travel from place to place quickly.

But we have to be very careful to use the technology we have in a responsible way. The same knowledge that saves lives and makes our world better can also lead to the development of powerful weapons and chemicals that can harm us and our planet.

Science and Machines
Energy and Industry
Transportation

SCIENCE AND MACHINES

You don't need to walk into a laboratory to see science in action. Science is all around you, every day of your life. There are three main kinds of science. The science of the natural world – plants and animals – is called biology. The study of what things are made of and what happens when they are mixed together is chemistry. The third science, physics, is about how everything in the universe works.

The world of physics includes the study of machines. Machines help us perform tasks with less effort. For example, one of the first machines, the wheel, makes it easier to move things from place to place. From scissors to smoke detectors, machines are all around us, making our lives easier and safer.

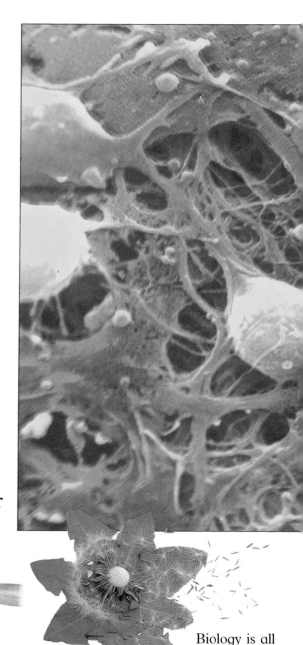

The laws of flight are studied in physics.

When you heat eggs, a chemical reaction makes the runny part hard and white.

Biology is all about living things.

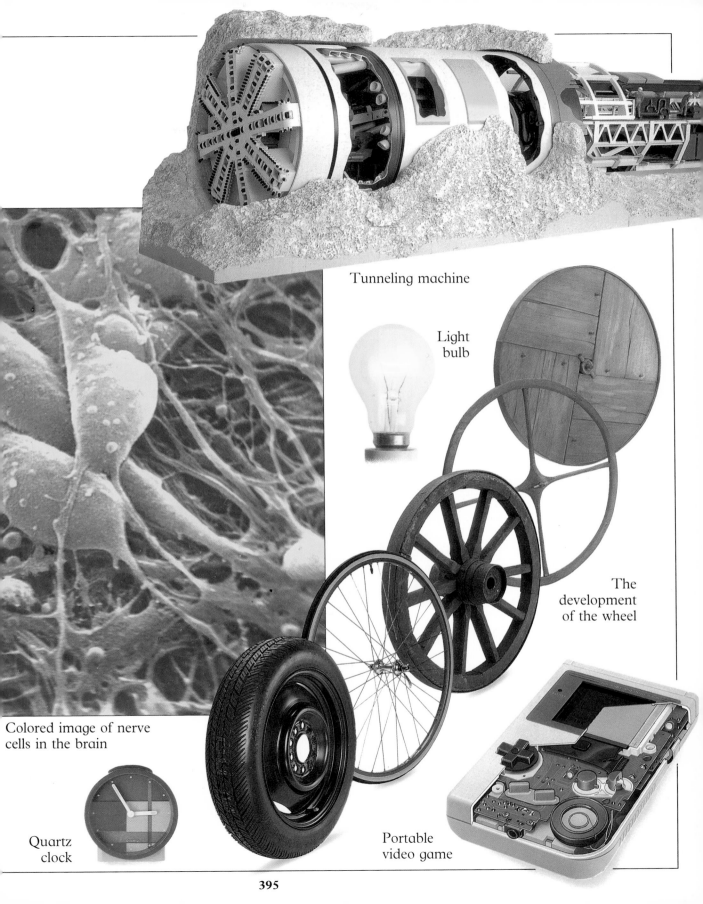

Tunneling machine

Light bulb

The development of the wheel

Colored image of nerve cells in the brain

Quartz clock

Portable video game

LIVING THINGS

Cherries

You are a living thing. All living things are made up of cells, and your body is made up of many billions of cells. Your cells join together to make tissues – for example, muscles. Tissues combine to make organs, like your heart. Biologists are scientists who study living things. They have divided the world of living things into groups, called kingdoms. The study of plants is called botany, and the study of animals is called zoology.

Seaweed

Moss

Fly agaric fungus

Bacteria and Fungi Kingdoms

Bacteria are so tiny, they can't be seen without a microscope. They have a simple, one-cell structure. Bacteria and fungi feed on other living things and recycle the remains.

A kitten grows up into a cat. All living things grow.

Splitting Up

Living things are made up of at least one cell. For a living thing to reproduce and grow, its cells must keep on splitting into identical pairs. This single-cell bacterium divides every 20 minutes. Look how many bacteria there are after one hour.

Now

20 minutes later

40 minutes later

One hour later

All living things make copies of themselves. This is reproduction.

Living things all find some way of breathing. A cat breathes through its nose.

Living things need food to stay alive.

Plant Kingdom

There are many different kinds of plants. Some are soft and small enough to hold in your hand. Others are enormous and woody. Some plants have flowers, but others don't. Green plants use sunlight and water to make their own food.

Sweet chestnut

Poppy heads and seeds

All living things get rid of waste products.

Gerbera flower

Raft spinner

Hoverflies

Badger

Animal Kingdom

Animals have billions of very complex cells. Most of the world's animals, like insects, do not have backbones. Animals with backbones include mammals, birds, reptiles, amphibians, and fish.

Hen with chicks

Grass snake

Spotted Salamander

Everything that lives, moves, but you can't always see the movement.

Lovely Lashes

Some people have tiny mites living in their eyelashes. The mites feed on liquid from the eye, cleaning the lashes at the same time. The mites need people, and people need them. Many living things depend on each other.

A cat hears with its ears. Every living thing is sensitive in some way to the world around it.

Cuban hock

Shore crab

Rudd

SOLIDS

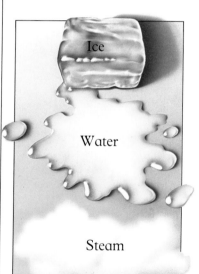

Everything is made up of matter. Molecules are the building blocks of matter. In solids, molecules are usually packed together in a regular way, so the solid keeps its shape without needing a container to hold it. Most matter is visible, like the pages of this book, but some is invisible, like air. Matter may be solid, liquid, or gaseous and change from one state to another.

Hydrogen atom

Uranium atom

Amazing Atoms
The smallest amount of a pure substance is called an atom. The simplest atom is hydrogen, but other atoms are more complex. Atoms are incredibly tiny – about 100 billion atoms fit on this period.

Ice

Water

Steam

Molecules in a Solid
Each of these children is acting the part of a molecule in a solid.

The children are all wearing red T-shirts to show that the molecules are all the same.

The molecules in most solids make up a very regular pattern.

In a State
Ice, water, and steam are made up of the same molecules. They are not alike, because their molecules are spaced differently. Solid ice melts into liquid water. When water is boiled, it turns into a gas, called steam.

Solids on Show
You can tell these solids apart because they do not look, feel, weigh, or smell the same. They all have different properties.

A flower is soft and delicate.

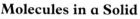

A muffin is crumbly and tasty.

Making Molecules

Atoms link up to make molecules. Two atoms of hydrogen join together to make one molecule of hydrogen gas.

Larger than Life

These long, thin wax molecules are shown about 3 million times larger than they are in real life. The tiny blue dots are atoms that have joined together to make up the wax molecules.

Sugar

Crystal Clear

Grains of sugar are solid crystals. Their atoms are close together and arranged in a regular pattern, called a lattice. Many solids that look smooth actually have a crystalline structure, like these vitamin C tablets.

The molecules in a cool solid do not move about very fast. If a solid is heated, the molecules move about faster and faster.

Vitamin C crystals, seen through a microscope

Vitamin C tablets

There are strong links between the molecules in a solid. This means that solids have a fixed shape.

A shell is hard and brittle.

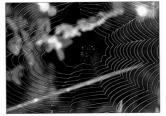

A spider's web is light and strong.

Graphite and Diamond

Graphite is soft and is used in pencils, and diamond is the hardest solid in the world. But diamond and graphite have a lot in common! They are two different forms of carbon. This means they have the same molecules, but they are arranged in different ways.

LIQUIDS

If you spread some butter on hot toast, the butter melts. When a solid gets hot enough to melt, it turns into a liquid. Liquids behave differently from solids. The heat that melts a solid breaks down some of the strong links between the molecules, so that the molecules can move about more freely. A liquid flows because its molecules can't hold together strongly enough to form a solid shape.

Syrup flows slowly.

Car oil flows quite well.

Ink flows very easily.

Sticky Spoonful
Some liquids flow much more easily than others. Liquids that are viscous do not flow well.

Molecules in a Liquid
These children are behaving like molecules in a liquid.

Liquid Levels
Liquids flow to fill containers of any shape or size. The surface of a liquid always stays level, however much you may tilt the container.

Fair Shares?
Who has more milk to drink? It may not look like it, but these glasses hold the same amount. Containers of different shapes can hold the same volume, or amount, of liquid.

Too Hot to Handle?

Ice cream melts easily, but not all solids have such low melting points. Some only melt if they are heated to very high temperatures. The steel in this factory melts at 2,700°F (1,500°C).

On the Boil

When a liquid is heated to a certain temperature, it turns into a gas. This temperature is called the boiling point. Different liquids have different boiling points. Water boils at 212°F (100°C).

In a liquid, the molecules are separate from each other.

Moving molecules can get right into all the corners of a container. This is why liquids take on the shape of the container that holds them.

Molecules in a liquid are not arranged in a regular pattern, so they can move about freely.

Molecules in a liquid move about faster than molecules in a solid.

On the Surface

Molecules near the surface of a liquid pull toward each other. A drop keeps its shape because of this surface tension. If some dish soap is added to a water drop, the surface tension is made weaker, so the drop spreads out.

Good Mixers

Cranberry juice mixes well with water. It dissolves completely.

Bad Mixers

Some liquids do not mix at all. Oil does not dissolve in water.

GASES

Gases are all around us and can't usually be seen or felt, but some can be smelled. All smells are molecules of gas mixed in the air. If you smell some tasty soup, you are actually sniffing in molecules of the soup. When you heat a liquid, it turns into a gas. The heat makes the molecules in the liquid move about faster and faster. The gas molecules fly off in all directions, spreading through the air.

Lighter than Air
When a gas gets hot, it takes up more room and gets lighter. A hot-air balloon rises up in the air because the air inside it is hot and much lighter than the air outside.

Molecules in a Gas
Each child is acting like a molecule in a gas. Look how they are bumping into each other.

There are no links between any of the molecules in a gas.

Dancing on Air
When a shaft of light shines through a gap in the trees, specks of dust look as if they are dancing in the sunlight. What is really happening is that the molecules in the air currents move about, flicking the dust in all directions.

As a gas spreads, the molecules get farther and farther apart. Gases will always spread to fit the space they are in.

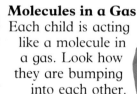

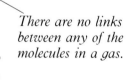

Carbon dioxide
Other gases
Oxygen
Nitrogen

In the Air
There are a number of different gases in the air we breathe. About 78 percent of air is nitrogen gas. Oxygen makes up 21 percent. Carbon dioxide and a variety of other gases make up just 1 percent of the gases in the air.

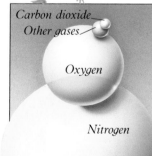

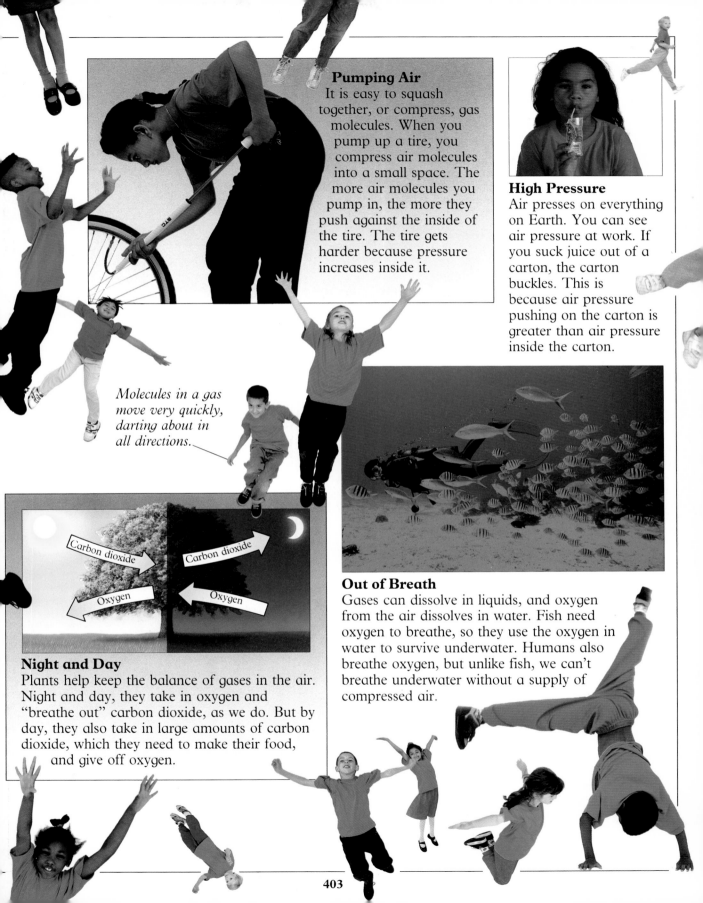

Pumping Air

It is easy to squash together, or compress, gas molecules. When you pump up a tire, you compress air molecules into a small space. The more air molecules you pump in, the more they push against the inside of the tire. The tire gets harder because pressure increases inside it.

High Pressure

Air presses on everything on Earth. You can see air pressure at work. If you suck juice out of a carton, the carton buckles. This is because air pressure pushing on the carton is greater than air pressure inside the carton.

Molecules in a gas move very quickly, darting about in all directions.

Night and Day

Plants help keep the balance of gases in the air. Night and day, they take in oxygen and "breathe out" carbon dioxide, as we do. But by day, they also take in large amounts of carbon dioxide, which they need to make their food, and give off oxygen.

Carbon dioxide

Carbon dioxide

Oxygen

Oxygen

Out of Breath

Gases can dissolve in liquids, and oxygen from the air dissolves in water. Fish need oxygen to breathe, so they use the oxygen in water to survive underwater. Humans also breathe oxygen, but unlike fish, we can't breathe underwater without a supply of compressed air.

ENERGY

Energy is needed for life and for every single movement in the whole universe. When you have run a race, you may feel that you have used up all your energy. But your energy has not been lost. It has changed into different kinds of energy: movement and heat. When work is done, energy is never lost, but it changes into other kinds of energy. Movement and heat energy are just two of the many different kinds of energy.

Jumping Jack
When the lid is closed, the puppet's spring is coiled up, ready to push the puppet out of the box. We say the spring has potential energy. When you open the lid, the spring's potential energy turns into movement energy.

Energy Changes
We can't make energy or destroy it. Instead, energy can change from one kind to another. This toy robot shows how energy may not stay in one form for very long!

Eating Energy
We get our energy from food, and we need to eat plants, or animals that have eaten plants, to stay alive. Plants get energy to grow from sunlight, so our energy really comes from the Sun.

Sun Power
Living things that grew millions of years ago were buried under rock, where they slowly turned into coal, oil, and gas. Energy from these fuels comes from the Sun, shining long ago.

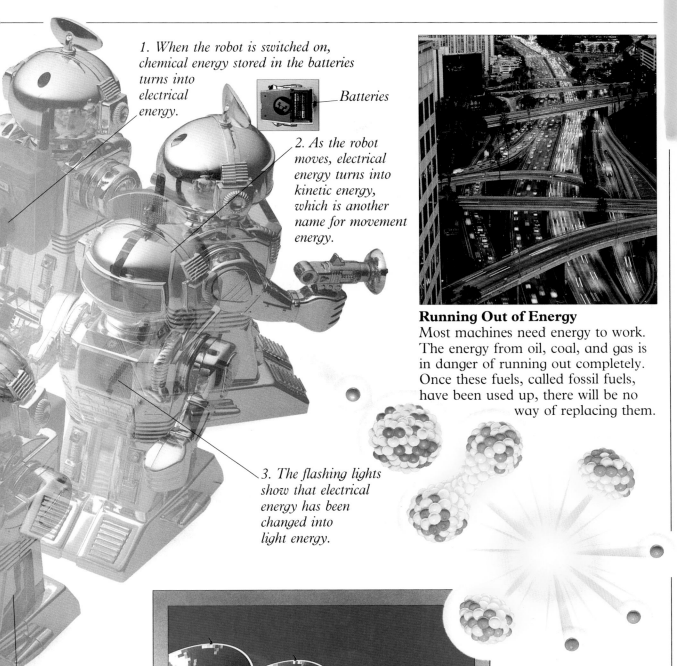

1. When the robot is switched on, chemical energy stored in the batteries turns into electrical energy.

Batteries

2. As the robot moves, electrical energy turns into kinetic energy, which is another name for movement energy.

3. The flashing lights show that electrical energy has been changed into light energy.

4. The robot makes a noise as it moves. Electrical energy has been turned into sound energy.

5. The robot gets warm. Movement energy has changed into heat energy.

Running Out of Energy
Most machines need energy to work. The energy from oil, coal, and gas is in danger of running out completely. Once these fuels, called fossil fuels, have been used up, there will be no way of replacing them.

Everlasting Energy
The Sun's rays beat down on these solar panels. Heat is stored and used to make electricity. Energy from the wind and waves also provides power. Scientists are trying to find cheaper ways of capturing the endless supply of natural energy.

Splitting the Atom
Neutrons are tiny particles inside an atom. If a uranium atom takes in one extra neutron, it splits in half, releasing a huge amount of energy. Nuclear energy produces radioactive waste, which has to be handled and got rid of very carefully.

HEAT

The Sun's rays can make you very hot. The way heat rays move through air is called radiation. You feel warm because the radiation makes the molecules in your skin move about faster than usual. Heat comes from molecules moving around.

It moves through solids by conduction – the molecules in a solid vibrate, bumping the heat along. Heat travels through air and liquids in a circular movement, called convection.

Heat moves along the wooden spoon by conduction. The spoon takes a long time to get hot because wood is not a good conductor of heat.

Snug as a Bug?

Heat doesn't travel easily through air, so materials that trap air keep you warm. This polar explorer is wearing clothes made of materials that insulate his body. This means they keep his body heat close to his body, where he needs it.

Tricky Fingers

If you dip one finger in hot water and another in cold, the hot water will feel hot and the cold water will feel cold. If you then dip both fingers in lukewarm water, your "hot" finger is tricked into finding the water cold, and your "cold" finger finds the water hot.

Good Conductors

If you heat a solid, its molecules jostle about, passing the heat from molecule to molecule. Just as a whisper moves along a line of children, so the moving molecules pass, or conduct, the heat from one end of the solid to the other.

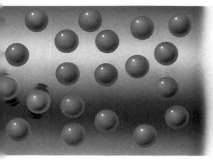

Building Bridges

When most solids are heated, they get bigger, or expand. This is because the heat speeds up the molecules, and they get farther and farther apart as they move. Bridges are built with small gaps between the long pieces of metal, so there is room for the metal to expand on a hot day.

As the water reaches its boiling point, it evaporates into steam.

When the water near the heat source gets hot, it rises up toward the surface.

The cooler water near the surface will sink to the bottom of the pan, where it is then warmed up by the heat source.

The flames are a direct source of heat.

Metal is a good conductor of heat.

Familiar Temperatures

212°F (100°C)
Water boils

140°F (60°C)
Cup of tea

98.6°F (37°C)
Body temperature

86°F (30°C)
Hot summer day

50°F (10°C)
Cold drink

32°F (0°C)
Water freezes

LIGHT

You may need to wear sunglasses on a sunny day because the bright light hurts your eyes. It is dangerous to look straight at the Sun. The Sun's light is not just blinding, it also travels to us very fast. In one second, light travels about 186,280 miles (300,000 km). If a straight beam of light has to pass through obstacles like convex and concave lenses, it changes direction. This is called refraction.

Hot Stuff
Most light comes from hot objects. A very hot object makes a very bright light. The Sun is white hot, and sunlight is the brightest natural light that we have.

Concave lenses curve inward in the middle.

This beam of light comes from a flashlight. A flashlight is a human-made light source.

Light bends outward when it passes through a concave lens.

Convex lenses bulge in the middle.

Retina

Pupil

Optic nerve

Lens

How We See Light
Light enters your eye through your pupil. It passes through a lens, which focuses the object you are looking at onto the retina. Millions of tiny cells inside the retina turn this upside-down image into electrical messages. The optic nerve carries these messages to your brain, which "sees" the image the right way up.

Light for Life
Plants always grow toward the light, even if, like this plant, they have to grow around corners to reach it! Light is very important for life. Plants need light to grow, and we need to eat plants, or animals that eat plants, to stay alive.

The Big Bang
Light travels faster than sound. When fireworks explode high in the sky, you see the lights before you hear the big bang. Light reaches your eyes more quickly than sound reaches your ears.

Over the Rainbow
Drops of rain act like tiny prisms. When light passes through raindrops, the colors of light are split up to form a rainbow.

When a beam of light passes through a prism, the colors split apart because each wavelength is bent a different amount.

Light travels in tiny waves. Light has a mixture of different wavelengths.

A prism is a solid, triangle-shaped piece of glass or plastic.

When light is reflected, the beams of light bounce off the mirror at the same angle as they hit it.

When light passes through a convex lens, it is bent, or refracted, inward.

When light hits an object that it can't travel through, like this mirror, a shadow forms behind the object where the light can't reach.

Busy Line
When you talk on the telephone, your voice is turned into laser light signals and sent down very thin fiberglass tubes called optical fibers. Up to 150,000 different conversations can be sent down just one of these optical fibers.

COLOR

Imagine waking up in a world without color. There would be no beautiful paintings or rainbows to look at and no bright shoes or clothes to wear. At night, when you switch off the light, all the bright colors around you suddenly disappear. This is because you cannot see the color of an object without light. White light can be split into all the colors of the rainbow – shades of red, orange, yellow, green, blue, indigo, and violet. When all these colors are mixed back together, they make white light again.

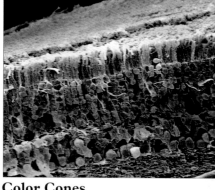

Color Cones

The inside of the back of your eye looks like this through a microscope. You have cells in the back of your eyes called cones, which send messages to your brain about the colors you see.

Seeing in Color

If you don't have all the different types of cones in your eyes, you may find it hard to tell some colors apart. A color-blind person may not be able to see the number on this color-blindness test.

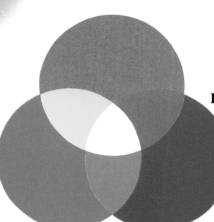

Seeing Red

We see things because light bounces off them. Light is actually a mixture of all the colors of the rainbow. When light hits these shoes, all the colors sink in, except the red color that is reflected back to your eyes.

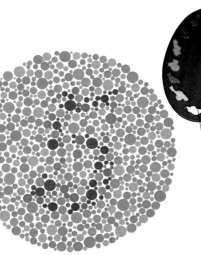

White is the only color that can't be made by mixing the primary colors of paint together.

Light Mixing

The colors of light behave differently from paint colors. When red, green, and blue light are mixed together, they make white light. The same colors of paint mix to make a dark color.

Paint Mixing

You can make new colors by mixing the primary colors of paint together.

Blue, yellow, and red are the primary colors of paint.

Red and yellow mix to make orange.

Blue and red mix to make purple.

Different shades can be made by adding more of one color than another.

Blue and yellow mix to make green.

Tasty Morsels

Which of these dinners would you like? Both meals would taste the same, but the food coloring in one turns you off before you start. Color can change our feelings about things, especially food.

Warning Colors

Colors are often used as a warning. This moth is poisonous, and its bright wings warn hungry birds to leave it alone. We also use colors to warn us of danger. A red traffic light tells us to stop, but a green traffic light means that it is safe to go.

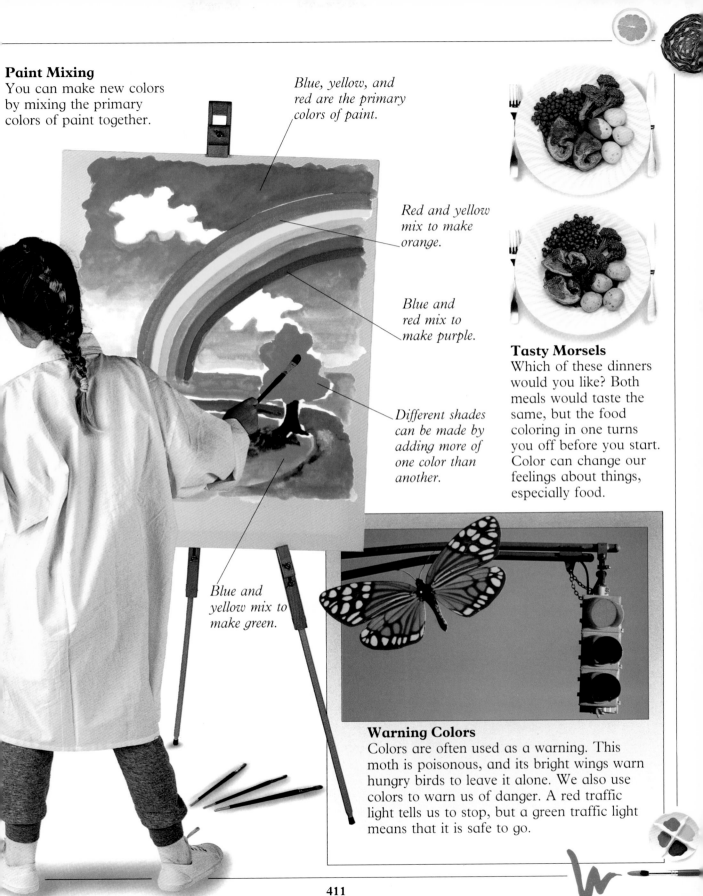

SOUND

Astronauts talk to each other by radio because their voices can't travel through empty space. Sound travels in waves. When these waves move through air, the air molecules move quickly, or vibrate. If there are no air molecules, no sound is made, because there is nothing to vibrate. When you shout, the vocal cords in your throat vibrate. These vibrations pass through your mouth into the air, making the air itself vibrate. Your ear picks up the vibrations, and you can hear them as sound.

Deafening Decibels
The loudness of sounds is measured in decibels. When airplanes land, they reach a very high number of decibels. This ground controller is wearing ear protectors to keep from being deafened by the noise.

Wailing Whales
Whales communicate underwater over huge distances. The sound of a big whale may travel hundreds of miles underwater. Sound travels faster and farther through liquids and solids than through air. This is because molecules in liquids and solids are more tightly packed.

The bottle tops clink together when the jingling stick is shaken.

Lentils inside the shaker make a soft, rattling sound.

Strings vibrate as the instrument is plucked. The tighter the string, the higher the note.

412

Seeing Sounds

You can show that sound vibrations exist by hitting a tray next to a drum sprinkled with rice. The rice bounces up and down with the vibrations that the sound makes.

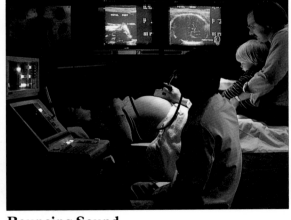

Bouncing Sound

This pregnant woman can see her baby before it is born. High-frequency sound waves are reflected as they reach the baby, making an image on the computer screen.

Hitting a metal plate makes it vibrate. You hear a loud and crashing noise.

What Is Frequency?

The number of complete sound waves that pass by in a second gives us the "frequency" of a sound. Frequency is measured in hertz. High notes have a high frequency. They make lots of vibrations and have a high number of hertz. Low notes have a low frequency.

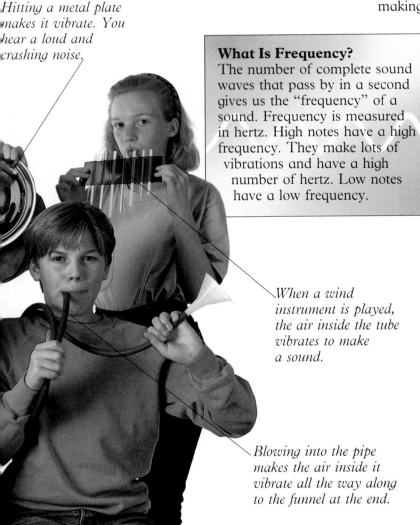

When a wind instrument is played, the air inside the tube vibrates to make a sound.

Blowing into the pipe makes the air inside it vibrate all the way along to the funnel at the end.

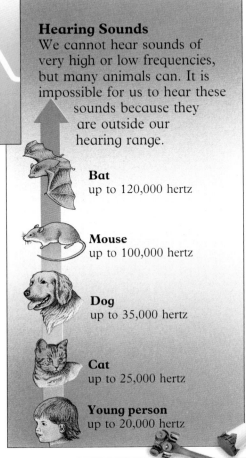

Hearing Sounds

We cannot hear sounds of very high or low frequencies, but many animals can. It is impossible for us to hear these sounds because they are outside our hearing range.

Bat
up to 120,000 hertz

Mouse
up to 100,000 hertz

Dog
up to 35,000 hertz

Cat
up to 25,000 hertz

Young person
up to 20,000 hertz

FORCE AND MOVEMENT

A rocket can only take off because a blast of energy forces it to move. Things move if they are pushed or pulled, or if they are high up to start with. The movement they make has a speed in a particular direction. How quickly they speed up depends on how big the force is and how heavy the moving thing is. Some forces make things move, but others stop things from moving. Friction is the name for the force that stops movement.

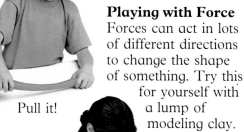

Push it!

Pull it!

Playing with Force
Forces can act in lots of different directions to change the shape of something. Try this for yourself with a lump of modeling clay.

Squeeze it!

Roll Apart
If one girl pushes the other, both will start to roll away from each other. When a force pushes one way, it also pushes equally in the opposite direction.

The go-cart wheels will keep moving until something stops them or slows them down.

If the passenger is heavy, it is hard to push the go-cart along and make it speed up.

Friction is a force that slows things down. The go-cart wheels rub against the ground, making friction.

Speed and Velocity

The time something takes to move a certain distance is called speed. Velocity is speed in a particular direction. A rocket's velocity is much greater than a snail's!

Down to Earth

When you jump, a strong force called gravity pulls you back down to Earth. Earth's gravity gets weaker as you travel away from it. On the Moon, gravity is much less strong, so astronauts can jump high with heavy packs on their backs before gravity pulls them back down again.

A snail moves at about 0.003 m/h (0.005 km/h).

A fast sprinter runs at about 22 m/h (36 km/h).

A race car reaches speeds of up to 210 m/h (338 km/h).

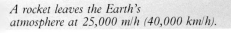

A rocket leaves the Earth's atmosphere at 25,000 m/h (40,000 km/h).

Pushing is a kind of force. The harder the boy pushes, the bigger the force on the go-cart.

Quick Stop

Without friction between its wheels and the ground, this truck could not move anywhere at all. The thick patterns on the tires, called tread, help the truck get a grip on the ground.

More pushing power is needed to start the go-cart than to keep it moving.

Good Performance

At high speeds, air rushes over the top of a race car, pushing it down onto the track. This makes the wide tires grip the track better, so the car can take turns faster than an ordinary car.

MAGNETS

Magnets are pieces of special material that have an invisible force that can push things away or pull things toward them. The biggest magnet of all is the world itself. Scientists think that as the Earth turns around, electricity is made in the hot metal deep down in the center. This electricity gives the Earth magnetic power. The Earth has two magnetic poles, called the magnetic North and South poles. Compass needles always point to the magnetic North Pole.

Magnetic Fields

A magnetic field is where a magnet has its power. If iron filings are sprinkled around a magnet, they gather together where the magnetic field is strongest. Two identical poles push each other away, so iron filings curve outward.

Two different poles attract. The iron filings run straight between them.

Left Out

Magnets only attract some kinds of material. Not all metals are attracted by magnets.

N
S
S
N

Magnets are fun to play with! These magnets are called horseshoe magnets because of their curved shape.

The area where the magnet has its power is called the magnetic field.

What Are Magnets?

Scientists think that atoms behave like tiny magnets. In a nonmagnetic material, the atoms face in different directions. In a magnet, all the atoms face the same way. Magnets lose their strength if they are hit, dropped, or heated, as this makes the atoms face in different directions.

Make a Magnet
If you stroke a needle about 50 times with a magnet, it will become magnetic. Stroking the needle in the same direction turns its atoms to face the same way.

Simple Compass
Compasses are made with magnets – see for yourself! Tape your magnetic needle to a piece of cork and float the cork in some water. It points the same way as a real compass needle – toward the magnetic North Pole.

The ends of a magnet are called its poles.

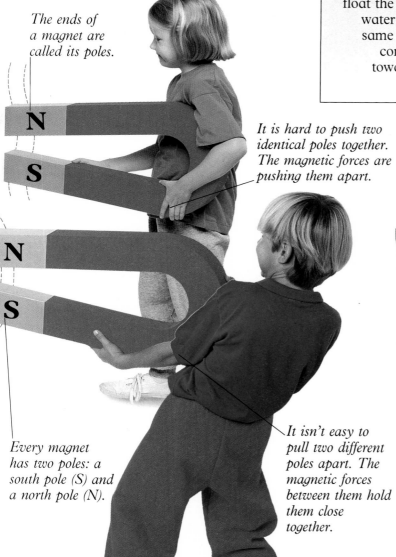

It is hard to push two identical poles together. The magnetic forces are pushing them apart.

Every magnet has two poles: a south pole (S) and a north pole (N).

It isn't easy to pull two different poles apart. The magnetic forces between them hold them close together.

Electricity and Magnets
Electricity is used to make magnets that can be switched on and off. This electromagnet is used to sort scrap metal. You can make an electromagnet by winding some wire around a nail and connecting it to a battery. The nail becomes magnetic.

MACHINES

When you do a difficult sum, you are working hard, but this is not what scientists call work! In science, work is only done when something is moved – for example, lifted or turned around. Machines make it easier to move things, so less effort is needed to do a job. Bicycles help us move quickly. Like most machines, bicycles are made up of a number of small, simple machines. Wheels, axles, pulleys, gears, levers, slopes, and screws are all common simple machines.

Forced Apart

Wedges are simple machines that split things open. When a force hammers a wedge into a block of wood, the wood is pushed apart. A wedge is a kind of slope.

Fulcrum

Saddle Sore?

It is possible to ride a long way on a bike – between 1922 and 1973 a Scottish man, Tommy Chambers, cycled 799,405 miles (1,286,517 km)!

Easy Does It

Levers make hard jobs easier. A seesaw is a kind of lever. The balancing point is called a fulcrum. The girl's weight is the force, and the box is the load.

Cables link the brake levers to the brake pads. Pulling the brake levers makes the brake pads stop the moving wheels.

Brake pads press against the wheel. They use friction to stop it from moving.

Wheel Genius

The wheel is one of our most important machines. Two wheels can be joined together by a pole, called an axle. A small movement of an axle makes a big turn of a wheel.

The wheel spins, or rotates, around an axle.

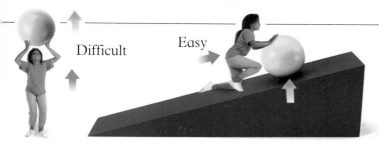

Difficult   Easy

Get into Gear

A toothed wheel that turns another toothed wheel is called a gear. Gears change the speed or direction of the moving part of a machine. The small wheel spins around twice as fast as the big wheel, turning in the opposite direction.

Uphill Climb

Slopes are simple machines. They make it easier to move things onto a higher or lower level. A screw is a kind of slope. If you could unwind the spiral grooves on a screw, they would flatten out as a slope.

A chain links the gears and the pedals to the back wheel. A small movement of the pedals is turned into a bigger turn of the wheel.

The chain is part of a pulley system.

Pedal power is the force that gets you started.

Light as a Feather?

Lifting up a weight with your bare hands is hard work. It is much easier to lift a weight by pulling down on the ropes of a pulley. If you double the number of pulley wheels, the same amount of effort will lift twice as many weights.

ENGINES AND MOTORS

Engines and motors provide power to make things move. Motors usually run on electricity and drive small things, like hair dryers. Engines are usually more powerful and run on heat. In steam engines – the first real engines – heat boils water to make steam, and the steam pushes the engine around, just as steam in a pan of boiling water pushes up the lid. Cars have "internal combustion" engines. In these, the engine is pushed around by pressure produced from burning gasoline inside the engine.

Boiler to make steam

Letting Off Steam
Until about 40 years ago, brightly colored steam traction engines like this one were often seen at fairgrounds. The traction engine drove the rides and made electricity for the lights.

Electric Power
Electric motors work by magnetism. An electric current passing through a coil of wire turns the coil into a very powerful magnet. If the coil is set between another magnet, it is driven around at great speed to power your hair dryer.

Magnet

The coil of wire spins around.

Jet Power
Most modern airplanes have jet engines. These push a blast of hot air out the back, which drives the plane forward at very high speed.

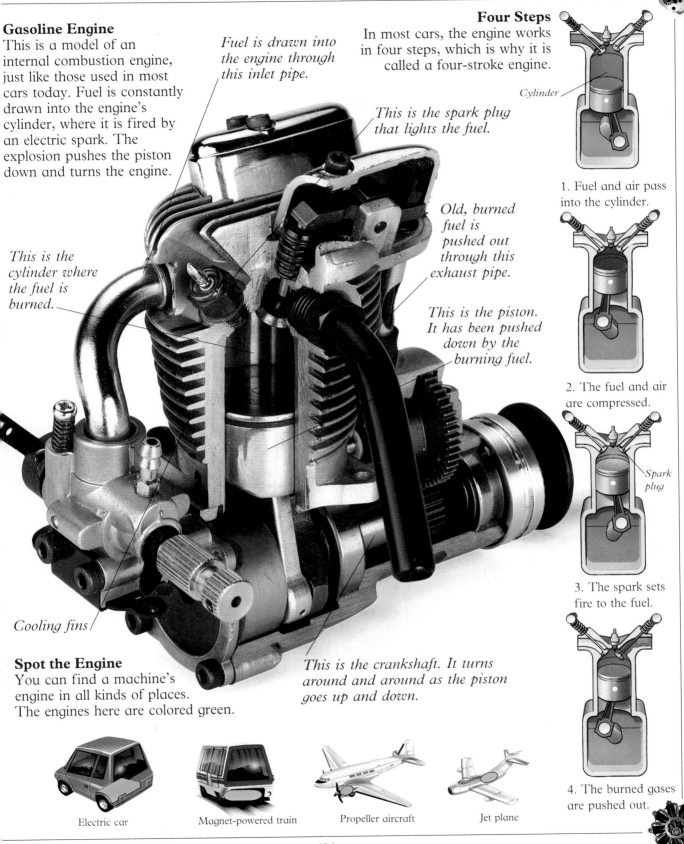

Gasoline Engine

This is a model of an internal combustion engine, just like those used in most cars today. Fuel is constantly drawn into the engine's cylinder, where it is fired by an electric spark. The explosion pushes the piston down and turns the engine.

Fuel is drawn into the engine through this inlet pipe.

This is the spark plug that lights the fuel.

This is the cylinder where the fuel is burned.

Old, burned fuel is pushed out through this exhaust pipe.

This is the piston. It has been pushed down by the burning fuel.

Cooling fins

Four Steps

In most cars, the engine works in four steps, which is why it is called a four-stroke engine.

Cylinder

1. Fuel and air pass into the cylinder.

2. The fuel and air are compressed.

Spark plug

3. The spark sets fire to the fuel.

4. The burned gases are pushed out.

Spot the Engine

You can find a machine's engine in all kinds of places. The engines here are colored green.

This is the crankshaft. It turns around and around as the piston goes up and down.

Electric car

Magnet-powered train

Propeller aircraft

Jet plane

WAR MACHINES

An aircraft carrier

Hand grenade

Going Bang!

Gunpowder was first used in China about 1,000 years ago for making fireworks.

In 1605, Guy Fawkes tried to blow up the English Parliament with gunpowder.

In 1867, Alfred Nobel invented a very powerful explosive, called dynamite.

Modern armies have all kinds of weapons, from rifles and grenades for fighting a single enemy, to bombs and missiles for attacking large targets. Soldiers can go into battle riding in a tank – a huge gun on wheels protected by heavy armor plating. But nothing can protect against nuclear bombs, the most powerful weapons of all.

The main gun can fire powerful shells that will travel 9,900 ft (3,000 m).

This is the turret. It turns around so the gun can be aimed in any direction.

The tank can carry a huge amount of fuel – around 500 gallons (1,900 liters). But it is only enough to drive about 270 miles (440 km) at 25 miles (40 km) an hour.

The commander sits at the top.

The driver sits inside at the front and uses mirrors to see what is happening outside.

The gunner aims the main gun using a thermal-imaging sight. This picks up the heat given off by enemy targets.

The loader loads the main gun with explosive shells and also operates the radio.

Runway at Sea

Aircraft carriers are huge ships with flat decks, which warplanes use for taking off and landing. The carriers sail close to enemy coasts so the planes can launch attacks.

One of the crew uses a machine gun to defend the tank against enemy aircraft.

Best Shot

Muskets were used by soldiers in the 17th century. They needed reloading after each shot.

Heavy steel armor up to 5 inches (13 cm) thick protects the crew from enemy fire.

Handguns like this Colt 45 were carried about 150 years ago by cowboys in the American west. They could fire six bullets without being reloaded.

Machine guns were invented in 1884 by Sir Hiram Maxim. They could fire dozens of bullets one after the other.

The wheels run inside tough metal bands, called caterpillar tracks. To steer, the driver makes one track run faster than the other.

Destruction caused by a nuclear bomb in Hiroshima, Japan, 1945

Nuclear Weapons

A single nuclear bomb can destroy a whole city, and the radiation it leaves behind can kill people and animals years later. There are now enough nuclear weapons to destroy the world.

A Nuclear Explosion

1. As the bomb explodes, it makes a giant fireball.

2. The explosion shakes the city below.

3. The blast and fire destroy buildings.

4. Rubble and dust mushroom up high into the sky.

COMPUTER MAGIC

Computers are the world's smartest machines. Inside a computer are thousands of very tiny electronic switches. By switching these on and off in different combinations, computers can perform all kinds of tasks. They are used every day in car factories, hospitals, supermarkets, and offices. Some guide aircraft, ships, submarines, and spacecraft. Other kinds, such as virtual-reality machines, are used for fun and can take you on exciting imaginary voyages.

The headset has a mini TV screen in front of each eye and a speaker over each ear.

When you move your head, this cable sends signals to the computer.

In a virtual-reality machine, you can imagine you are at the controls of a jet, a spacecraft, or even a mechanical dinosaur.

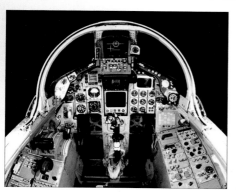

Keeping Your Head Up
Jet fighter pilots must not take their eyes off the view ahead, even for a second. All the information a pilot needs to control the plane and fire at targets is fed to a computer and projected onto the pilot's face mask. This is called a "head-up" display.

Feeling the Way
Doctors can use virtual-reality gloves to look inside you before they decide to operate. As they run their hands over you, sensors in the gloves send signals to a computer. The computer makes a 3-D picture of your insides on a little TV screen in the doctors' headsets.

This cable carries signals from the computer for the pictures and sounds in the headset.

When you turn your head, different views come up on the TV screens in the headset.

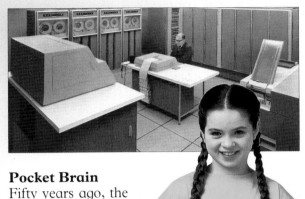

Pocket Brain

Fifty years ago, the first computers filled a large room. Now an equally clever computer can be the same size as a pocket calculator.

Bytes and Megabytes

Computers can help with homework or get an astronaut to the Moon.

Calculator

Laptop computer

Shrinking Switches

The switches in computers have got steadily smaller and more complicated.

The first computers had big glass valves.

Personal computer

Computers now have tiny microchips.

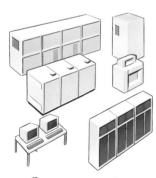

Super computer

Electric motors rock the virtual-reality machine backward and forward and from side to side to make your "journey" more realistic.

ROBOTS

Robots are machines that "think" with a computer brain that tells them what to do. Once they have been programmed, they can work entirely by themselves. Some people believe that one day we will be able to make robots that can do everything a human can – and they may even look like humans. At the moment, though, most robots are nothing more than mechanical arms or cranes.

Mechanical Humans

Automatons are machines that move rather like humans or animals. This one was made for a fair in Victorian times.

Factory Hand

A factory robot is often just a moving arm. But a robotic arm can hold things, screw them into place, weld them, and check that they work. It can replace lots of human workers.

An electronic voice allows the robot to answer and ask simple questions. It can also obey some clearly spoken commands.

Robots at Work and Play

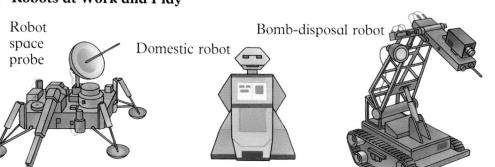

Robot space probe

Domestic robot

Bomb-disposal robot

Robotic arm

The robot can pick up your newspaper by squeezing its claws together. Its "muscles" are powered by electric motors.

Each of the claws has special pressure sensors so that they do not crush things they pick up.

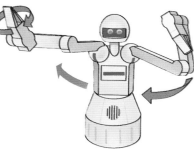

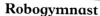

Robogymnast
The robot can turn its hands and move its arms. But unlike a human, it can turn its arms completely around.

Electronic eyes allow the robot to spot obstacles in its path and steer around them.

It can also turn at the waist or swivel on its base.

The robot's brain is a powerful computer. It tells the robot what to do by sending electronic signals to the motors that move the robot's different parts.

Home Help
A robot like this is really just a clever toy. But one day there may be robots that can do the shopping, cleaning, and other household chores for you.

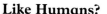

Like Humans?
Manny is a lifelike robot with computers that can make it "sweat" and "breathe." Scientists use Manny to test special clothing, like space suits and fire-fighting clothes. If Manny can "breathe" and doesn't "sweat" too much, then the clothes will be safe for a human to wear.

Manny has joints that move just like a human's.

ENERGY AND INDUSTRY

In today's world there are thousands of different industries, and they are divided into three groups. Primary industries, such as farming and mining, take raw materials from the Earth. Manufacturing industries make, or manufacture, useful things from the raw materials. Service industries are made up of people who sell these goods or supply skills such as nursing or teaching. All industry needs energy to work, so as the world becomes more industrial, it needs more and more energy.

Energy for industry comes mostly from fossil fuels – coal, oil, and gas. These are being used up quickly, and soon they may be gone.

Oil rig

Wind is a type of endless energy.

Oil

Coal

Natural gas

Nuclear energy

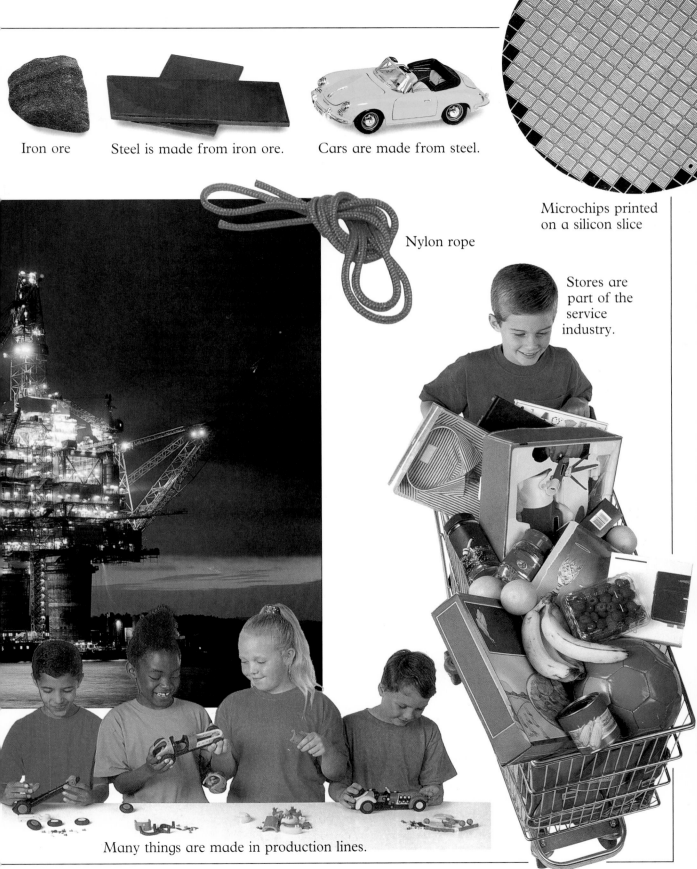

Iron ore

Steel is made from iron ore.

Cars are made from steel.

Microchips printed on a silicon slice

Nylon rope

Stores are part of the service industry.

Many things are made in production lines.

COAL

Like layers of cream in a cake, there are bands, or seams, of coal in the rocks beneath the ground. Some seams are near enough to the surface to be scraped out with diggers. This is called strip mining. But most seams are found deeper down and have to be dug up by miners using massive machines. About half of the 5 billion tons of coal mined each year is burned in power stations to generate electricity.

Big wheels, called winding gear, lift the coal bucket to the surface.

Strip mine

Fans suck stale air out of the mine shaft. This drags fresh air down the other shaft.

Scoop

Cut It Out
Each scoop on this coal-cutting machine is big enough to hold a car!

Roof supports

Cooking Coke
Coal is put into giant ovens and baked at more than 1,650°F (900°C). When the oven door is opened, coke topples out. Coke is needed to make steel.

Coke is coal minus tar, oils, and gases.

The tar that collects at the bottom of a coke oven is used to make soap!

Face Facts
The part of a seam where coal is being cut is called a face. The coal is ripped out by a spinning cutter with steel teeth.

Coal face

A big bucket carries coal to the surface.

Water is sprayed onto the coal to cut down the coal dust.

Conveyor belt

Pumps take water from the bottom of the mine.

Air intake

Computer control room

The coal is washed and separated into different-sized lumps before it leaves the mine.

Burning to Go
Nonstop trains take coal straight to the power station.

Cages carry miners down a deep shaft.

The main shaft was blown out of the rock with explosives.

All Aboard!
Miners travel on trains to coal faces that are many miles away from the shaft.

Most of the coal in this seam has already been dug out.

Long tunnels are dug to reach new coal seams.

A breeze can always be felt in the tunnel as the air moves through the mine.

Boring!
Big machines chew through the rock to open up new tunnels.

Mine shafts can be 3,950 feet (1,200 meters) deep.

Down to Earth
Not all coal is the same. Hard coals, which are found deeper underground, release more energy when they burn.

Half-squashed coal, or peat, is made into fuel bricks.

Soft, crumbly brown coal is burned to make coke.

Black bituminous coal is burned to generate electricity.

Anthracite coal is used in houses and factories.

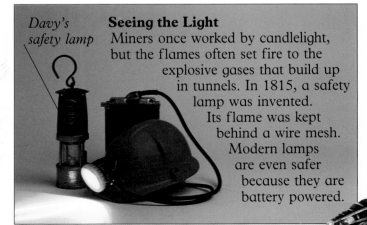

Davy's safety lamp

Seeing the Light
Miners once worked by candlelight, but the flames often set fire to the explosive gases that build up in tunnels. In 1815, a safety lamp was invented. Its flame was kept behind a wire mesh. Modern lamps are even safer because they are battery powered.

OIL

In the rocks under hot deserts, snowy plains, and stormy seas, there is buried treasure: a "liquid gold," called oil. Most of this sticky, black fossil fuel is used for energy, but 12 percent of each barrel is turned into chemicals and plastics. All oil is brought to the surface by drilling deep holes, called wells. On land this is fairly easy, but at sea, platforms as tall as skyscrapers have to be built.

Sandstone

Sandstone with oil

Solid "Sponge"
Oil is found in the tiny spaces in rocks such as sandstone. This oily layer is often sandwiched between water and a layer of natural gas.

Each marble stands for a grain of rock.

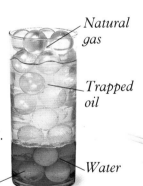

Natural gas

Trapped oil

Water

A Boring Bit
A sharp-toothed metal cutter, called a bit, bores through rock to reach oil. Drill bits are replaced twice a day because they wear out very quickly.

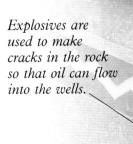

During drilling, chemicals are pumped around the bit to carry away rubble.

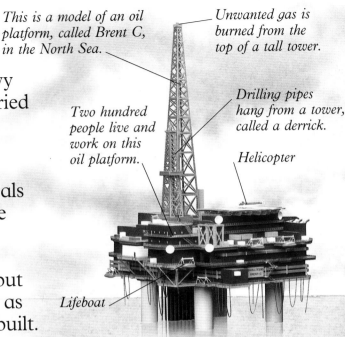

This is a model of an oil platform, called Brent C, in the North Sea.

Unwanted gas is burned from the top of a tall tower.

Two hundred people live and work on this oil platform.

Drilling pipes hang from a tower, called a derrick.

Helicopter

Lifeboat

Every day, one-quarter of a million barrels of hot, freshly drilled oil are pumped into the hollow concrete legs to cool down.

Some oil is piped into a gigantic, underwater storage tank.

Cooled oil is piped ashore.

Explosives are used to make cracks in the rock so that oil can flow into the wells.

Gas

An arched layer of oil-proof rock, such as granite, traps oil underneath it.

Oil

Water

Wells fan out to reach the oil.

Only Way Out

A pipe 797 miles (1,284 kilometers) long snakes across the snowy lands of Alaska. Oil takes a whole week to flow down the pipe to reach an ice-free port in the south.

Oil at Sea

Huge structures, called rigs, drill down to find oil. At sea, some rigs float on the surface, but others stand on the seabed.

Huge tankers "plug into" this oil store to take on oil.

Tankers can take oil all over the world.

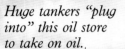

Look how big this rig is compared to the Statue of Liberty!

Jack-up rig

Anchors keep the oil store on the seabed.

Pump It Up

Not all oil gushes to the surface naturally. Some is pumped up by machines, called nodding donkeys. This "donkey" has been painted to look like a grasshopper!

Drill ship

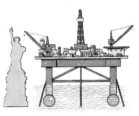

Semisubmersible rig

This zigzag break in the picture is to show that oil is usually found under thousands of feet of rock.

Gas gushes out when the pressure is released.

Wildcat Wanted?

Before a well is dug, geologists must be sure that the rocks below the ground are the right shape to trap oil. A test drill, or "wildcat," is only started if the surveys and satellite pictures look good.

Bubble Trouble

Just like the fizz in a fizzy drink, bubbles of natural gas are trapped in oil. If there is enough gas, it is piped ashore. If not, the gas is just burned.

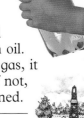

x

NATURAL GAS

What a Whiff!
Natural gas
has no smell.
Chemicals are
added to it so
that leaks can
be smelled.

In 1918, a gas was discovered in an oilfield in Texas. It was named natural gas because it replaced a gas that was manufactured from coal. This new fuel is now used in factories and homes across the world. Natural gas travels a long way before it reaches the top of your stove to burn as a bright blue flame. It has to be released from deep below the ground, cleaned, and piped countless miles.

The journey begins at a gas rig.

Gas terminal

Mostly Methane
Natural gas contains three different gases. Butane and propane are taken out at a gas terminal. Methane, the part that burns best, is sent through pipes to houses and factories.

Giant fans waft natural gas along the pipes.

Pipe

Butane gas camping stove

"Pig"

Soil

Cool It
Ships take methane to places that are not connected to pipelines. The gas is cooled into a liquid so that it takes up 600 times less room.

If it is cooled into a liquid, a balloonful of methane gas can fit into a space the size of a pea.

Methane is cooled to -260°F (-162°C) to make it turn into a liquid.

Methane tank

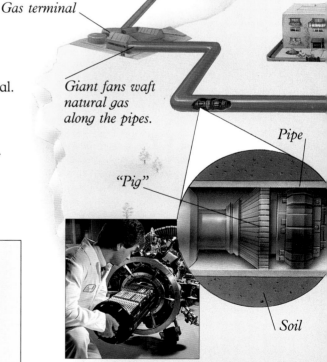

Very Important Pig
"Pigs," not people, check natural gas pipes! A "pig" is a computer on wheels that whizzes down pipes to pinpoint cracks and other problems.

On the Way Up

This big building, called a rig, gathers up gas that flows from deep under the seabed. The drill to reach the pocket of gas may be four miles (six kilometers) long!

What Size Pipe?

Some gas is stored near homes to supply sudden daily demands – such as at dinnertime!

The roof floats on top of the gas. So the lower the roof is, the less gas is left.

The plastic gas pipe is dragged through the tunnel by the "mole."

You could stand up in the pipe that travels between the rig and the terminal.

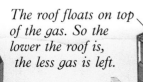

Iron pipe

Road

The "mole" smashes through the earth like a pneumatic drill.

Gas is sent through pipes to homes and factories.

A dog could fit into the pipe that links factory pipes to gas-terminal pipes.

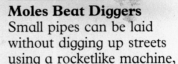

It is roomy enough for a cat to sit inside the pipes that take gas to factories.

Pumping stations keep the gas moving.

"Mole" hole

Growing Gas

Farms off the coast of California grow a giant seaweed, called kelp. It is harvested three times a year by special ships and then put in tanks and left to rot. The decaying kelp gives off methane gas.

Moles Beat Diggers

Small pipes can be laid without digging up streets using a rocketlike machine, called a "mole." Its route is guided by a computer.

A mouse could fit into the small plastic pipe that goes into your home.

NUCLEAR ENERGY

Not So Fast
This drawing shows very simply how neutrons whiz around a nuclear reactor and crash into uranium atoms in fuel rods.

Super Fuel
One handful of pure uranium can release as much energy as 72,000 barrels of oil!

Nuclear energy comes from atoms, the tiny particles that make up the whole universe. Enormous amounts of energy are locked inside atoms. When billions of uranium atoms are torn apart in a nuclear power station, the energy that is set free can boil water. Steam from this hot water is used to generate electricity. People worry about nuclear power because when the energy is released from an atom, deadly rays, called radiation, also escape.

2. The splitting uranium atoms inside the fuel rod "shoot out" new neutrons, which travel at 9,900 miles (16,000 kilometers) per second.

1. Energy is released when neutrons hit atoms in the fuel rods.

Water warms up

Fuel rod

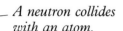
A neutron collides with an atom.

The center of the uranium atom splits in half.

Energy and radiation

Two or three new neutrons escape. Each one can collide with another atom and set free energy.

Old fuel rods are radioactive "trash."

Cold water

Fission Division
The heart of an atom, called a nucleus, is made up of neutrons and protons. These are held together by energy. When an atom is split, some of this energy is set free. Splitting an atom to release energy is called fission.

Cool It
Fuel rods are replaced every few years. Before the reusable uranium can be removed from them, the rods are cooled in a special pond.

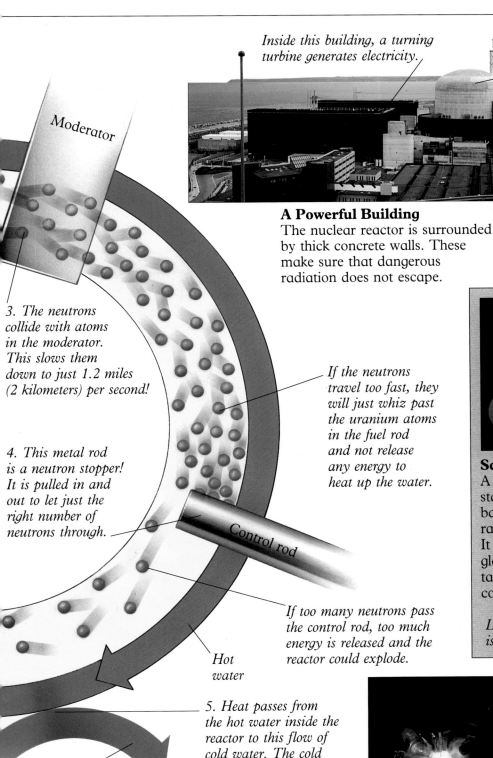

Moderator

Inside this building, a turning turbine generates electricity.

The nuclear reactor is in here.

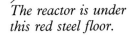

A Powerful Building
The nuclear reactor is surrounded by thick concrete walls. These make sure that dangerous radiation does not escape.

The reactor is under this red steel floor.

3. The neutrons collide with atoms in the moderator. This slows them down to just 1.2 miles (2 kilometers) per second!

If the neutrons travel too fast, they will just whiz past the uranium atoms in the fuel rod and not release any energy to heat up the water.

4. This metal rod is a neutron stopper! It is pulled in and out to let just the right number of neutrons through.

Control rod

Safe Deposit?
A typical nuclear power station produces about 20 bathfuls of very dangerous radioactive waste each year. It is made into a sort of glass and poured into steel tanks, which are coated in concrete and buried.

Less dangerous waste is buried in barrels.

If too many neutrons pass the control rod, too much energy is released and the reactor could explode.

Hot water

5. Heat passes from the hot water inside the reactor to this flow of cold water. The cold water boils into steam.

Laser beam

Sunny Future
When super-hot atoms collide, they fuse together and set energy free. It is this fusion that makes the Sun shine. Scientists are trying to build "Suns" on Earth by firing lasers at atoms.

6. The powerful jet of steam turns a turbine to generate electricity.

ENDLESS ENERGY

When oil, gas, and coal run out, people will need other sources of energy to fuel their cars and light their houses. Wind and water are already being put to work, but the best hope for an endless supply of free energy is the Sun. Light and heat from the Sun pour down onto the Earth all the time. Today, sunshine runs calculators, watches, and even power stations. One day scientists hope to collect sunlight in space and beam it back to Earth!

What a Gas!
In some countries, manure is collected, tipped into containers, and left to rot. The gas it gives off is piped to homes and used for cooking and heating.

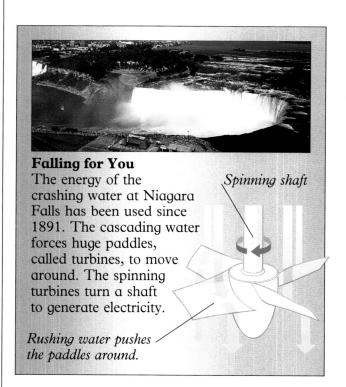

Falling for You
The energy of the crashing water at Niagara Falls has been used since 1891. The cascading water forces huge paddles, called turbines, to move around. The spinning turbines turn a shaft to generate electricity.

Spinning shaft

Rushing water pushes the paddles around.

Trick of the Light
This is the world's first solar power station. It was built in 1969 at Odeillo in France. Electricity is generated by using reflected sunlight to boil water into steam.

This enormous mirror is curved so that all the sunshine that hits it is reflected onto one small spot at the top of the tower.

Computers keep the 63 mini-mirrors facing the Sun.

Not Alone
The solar power station is faced by 63 small, flat mirrors. They reflect extra light onto the main mirror.

Whirling in the Wind
Strong, steady winds can be put to work turning windmill blades. As the blades spin, they turn a shaft, which generates electricity. These modern windmills come in several shapes. Groups of them are called wind farms.

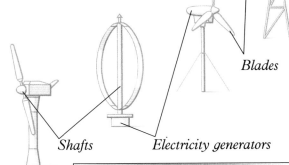

Blades

Shafts *Electricity generators*

Reflection of the ground

A solar power station does not need a chimney – there are no fumes or ash!

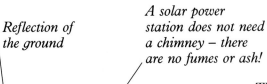

This mirror is 139 feet (42 meters) wide. It is built onto the side of a building.

Water inside this tower turns to steam.

It can get as hot as 6,870°F (3,800°C) inside this tower.

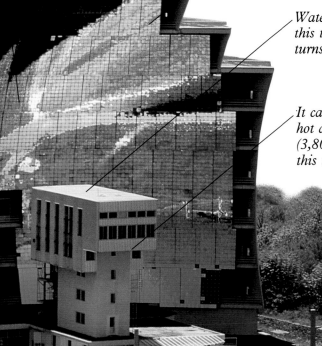

No Need for a Plug
Solar cells, made from slices of wafer-thin silicon, turn sunlight into electricity. This remote jungle telephone is powered by several solar cells.

Sunlight knocks electrons from the top layer to the bottom layer of silicon. This generates an electric current that is collected by the metal layers.

Sunlight
Metal
Silicon
Metal

ELECTRICITY

Electricity is used as a way of moving energy from place to place. It can take energy from burning coal in a power station into your home to work your television. Most electricity is generated in machines. Small machines, called dynamos, light the lights on bikes. Huge generators in power stations light whole cities. Pedal power works a dynamo, but steam produces the electricity in a power station. This steam is made by using the heat from burning fossil fuels or splitting atoms. Sunshine, falling water, and whirling windmills can also generate electricity.

Dynamo

Mighty Machine
The big blue generator inside this power station is about ten times as tall as you!

This magnet spins around because it is attached to a rod that touches the turning wheel.

Electricity is generated in this coil of wire by the spinning magnet.

Chimney

3. Steam surges from the boiler into the generator. It pushes around huge paddles, called turbines.

2. The burning coal makes water turn into steam.

1. Coal is crushed and then blown into the boiler to burn.

A pile of coal as heavy as 40 elephants is burned each hour.

4. Turbines turn a massive magnet around 50 times a second.

5. The moving magnet creates an electric current in huge coils of wire.

First transformer

A condenser turns the hot steam into hot water.

The cooling tower cools the hot water so that it can be used again.

Choice of Fuels
A power station uses just one sort of fuel to generate its electricity. This one burns coal, but other power stations use oil, natural gas, or nuclear fuel.

Oil

Natural gas

Uranium

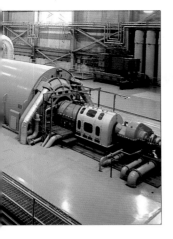

In Charge
Power stations can't be built near all the places that need electricity. So the electricity generated flows into a network of cables, called a grid. At the touch of a button, electricity is made to flow to wherever it is needed.

Electricity is dangerous, so tall pylons hold the long cables high above the ground.

Second transformer

Electricity cables are sometimes laid under the ground in towns and cities.

A substation makes electricity safe for you to use in your home.

Aluminum cable

Energy travels down the cables at about 155,000 miles (250,000 km) per second – almost as fast as the speed of light!

Going Up or Going Down?
Machines, called transformers, change the strength of an electric current. The current that flows between pylons has to be decreased to keep the cables from melting.

Watt Is Power?
The speed at which different machines use energy is measured in units, called watts.

Electric clock
(10 watts)

Vacuum cleaner
(1,000 watts)

Welding machine
(10,000 watts)

These children are pretending to be tiny parts of atoms, called electrons.

Each ball is a "parcel" of electrical energy.

Pass the Parcel
People once thought that electricity flowed like water, which is why it was called a current. In fact, energy moves along a cable more like balls being passed down a line!

METALS

Iron ore

Metals are found in the ground hidden in special rocks, called ores. Tin, copper, and iron all have to be taken out of their ores before a factory can melt and shape them into a can, pan, or car. Pure metals, however, are usually too weak to be used in industry and have to be mixed together to make better metals, called alloys. Lead is soft, and tin breaks easily. Together they can make a strong, tough alloy known as pewter.

Copper pan

Not Natural
Brass is an alloy, which means it can't be dug out of the ground. It is made by mixing two weaker metals: copper and zinc.

Brass is stronger than zinc or copper.

Zinc-coated bucket and wire

Precious Metals
Gold and silver are used to make far more than just jewelry. Gold is sprayed onto an astronaut's visor to reflect sunlight. Silver mixtures are used to make films because they are sensitive to light.

Steel from Iron
Iron has a lot of carbon in it, which makes it crack easily. If some carbon is removed, iron turns into super-strong steel. This change starts in a blast furnace.

Coke, limestone, and iron ore heat up and turn into iron and a waste material, called slag.

Limestone

Coke

Crushed iron ore

Blast furnace

Hot air

Slag

Iron

1. Blasted Iron Ore
A stream of iron is flowing from this huge oven, called a blast furnace. It has been burned out of iron ore by blasts of hot air.

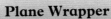

Big Dipper
Steel girders are dipped in a bath of melted zinc to keep them from rusting. This process is known as galvanizing.

Plane Wrapper
Aluminum is a marvelous metal. Thin sheets are wrapped around chocolate to preserve its taste. Thick sheets are made into jumbo jets. Aluminum is used to make planes because it does not rust and is very light. The aluminum is made as stiff as steel by adding a little copper.

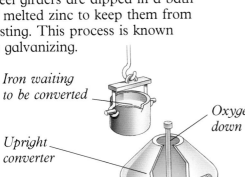

Iron waiting to be converted

Oxygen rushes down this tube.

Upright converter

Scrap iron can be put into the converter, too.

In 40 minutes the converter can make 350 tons of steel.

Rust Buster
You would not want to eat with rusty cutlery! So chromium is added to steel to make an alloy, called stainless steel.

Tipped-up converter

When the strips of steel are cold, they can be squashed by rollers into flat slabs.

Hot steel

2. Iron In, Steel Out
The liquid iron is poured into a converter. After a powerful jet of oxygen has burned out the impurities and most of the carbon, the converter is tipped up to pour out steel.

3. Taking Shape
Hot, freshly made steel is poured into a big tray. When it has almost set, nozzles are opened and steel oozes out, like toothpaste out of a tube.

MAKING A CAR

Model T

Every few seconds, somewhere in the world, a brand-new car rolls off a production line and out of a car factory. Each car is made from raw materials, such as iron ore, sulfur, and sand, which have been shaped into more than 30,000 parts! Most of this "jigsaw puzzle" is put together on a kind of giant conveyor belt. Each area of the factory puts on a few particular pieces; for example, the body shop adds the roof, but never the seats. The first car made like this was the Model T Ford.

Stamp It Out
Sheets of cold steel are stamped into shape by machines, called presses. A press room can be the size of three football stadiums!

Each car body is made from more than 20 pieces, or panels, of steel.

Press

Steel

The start of the "conveyor belt."

About 80 percent of a car is made from iron and steel.

Start with Steel
Steel is the most important ingredient for making cars. Rolling mills press hot steel into thin sheets. These are then rolled up and sent to car factories.

A roller test is used to check that the car is working.

Apart from their color, all the cars made on this production line are the same.

Cars are washed and polished before they leave the factory to be sold.

Ready to Go
Cars were only invented about 100 years ago, but there are now more than 400 million of them.

Built to Bounce
To make your journey smooth, tires are made of rubber and filled with air.

Strands of steel or nylon toughen tires.

Robot welder

The pressed steel panels form a rigid box to protect the passengers.

Robots in Charge

The steel sides, roof, and doors have to be joined together. This is done by welding – making the metal melt and stick together. Using more than 1,000 welds, robots can build a car body in just 42 seconds.

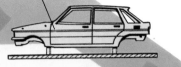

Robot painter

Mechanical Monets

Cars are painted by robots with sprays. The robots are not harmed by paint fumes and can put paint on quickly and accurately.

Heavy cars use more fuel. So more and more parts, such as bumpers, are now being made out of lighter materials, such as plastic.

The doors are removed so that the inside can be reached more easily.

Bare steel base *Top coat*

Fifteen coats of paint are put on each panel.

The car is lowered onto an engine, which was built on a separate production line. This may have been in another country!

Robots add windows.

Each worker repeats the same job over and over again.

Pile Up

A new car may not be as new as you think. Up to 40 percent of the steel may have come from old cars! Recycling scrap steel saves raw materials and energy.

Together at Last

The engine is the heart of the car, but it is not added to the body until near the end of the production line.

CHEMICAL INDUSTRIES

Oil refinery

Soap, fertilizer, and glue are just a few of the useful products of the chemical industries. They are made by combining different substances. Crude oil is the main raw material for these industries. The carbon and hydrogen in oil can be made to join up in different ways to make more than half a million things, such as paint or pills! This manufacturing starts in special factories, called refineries.

Made to Spread
Runny vegetable oil is made into solid margarine by adding hydrogen atoms.

Split It Up!
Crude oil is split into useful oils inside a distillation tower at a refinery. The oils are separated by being boiled into a gas and then cooled back into a liquid.

Liquid Again
If you put a saucer over a cup of steaming-hot liquid, droplets collect on the saucer. The liquid has turned into steam and then cooled back into a liquid when it hit the cold saucer.

Very hot liquid turns into a gas.

Gas cools into a liquid.

Distillation tower

230°F (110°C)

Cooling oil drips from the edge of this "saucer" into the tray below.

Kerosene is the fuel used by airplanes.

356°F (180°C)

The cloud of crude oil gets cooler and cooler as it wafts up the tower.

Lubricating oil makes machines run smoothly.

At about 725°F (385°C), crude oil changes into a gas.

Crude oil is pumped into a furnace to be boiled into a gas.

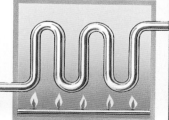

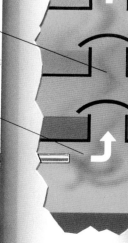

All Change

Chlorine keeps the water in swimming pools clean and safe to swim in. It is made in a factory by passing electricity through salty water. The electric current makes the atoms in the salt and water rearrange and produce chlorine.

Cat Cracker

Oil is made up of long chains of carbon and hydrogen atoms. Useful chemicals, called petrochemicals, are made by breaking up these chains. This is done by heating the oil in tanks, called cat crackers. The small chains can be used to make useful things such as shampoo.

Carbon atom   *Hydrogen atom*

Hydrogen atoms are forced between the carbon atoms to break up the chain.

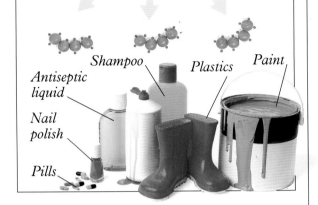

Shampoo   Plastics   Paint

Antiseptic liquid

Nail polish

Pills

The gases that come out of the top of the tower are made into plastics.

One-fifth of each barrel of crude oil separates into gasoline.

Gasoline is the most common fuel for automobiles.

When the cloud of crude oil reaches this height, it is cool enough for diesel to turn into a liquid.

Diesel is the fuel used by many trains.

A different oil flows out of each pipe because all the oils in crude oil cool into liquids at different temperatures.

Dark industrial oil is burned in factories and power stations.

Bitumen is the first oil to flow out of the tower.

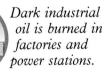

Thick, sticky bitumen is spread on the surface of roads.

Sulfur, So Good

About 150 million tons of sulfuric acid are used every year to help make things such as fertilizer, paper, and explosives! This acid is made by heating a yellow rock, called sulfur.

PLASTICS

Plastics are amazing materials. They don't rot like wood or rust like some metals. They are light and easy to shape. Plastic pens, shoes, and even surfboards are all made from oil or coal. Chemicals are taken from these fossil fuels and turned into small, white pellets. These are then melted and blown to form bags or rolled flat to make floor tiles. Bowls, boxes, and buckets are usually shaped by being injected into molding machines.

Get Set or Go?
Some plastics, such as Bakelite, are like bread! Once they have been "baked," they cannot be reheated and made into new shapes. A melamine mug will not change shape when hot drinks are poured into it. Polystyrene and polythene are more like chocolate – they can be melted again and again. Each time the mixture cools, it sets into the shape of the mold it has been poured into.

The two halves of the steel mold lock tightly together.

Hot Tub
Plastic tubs are made by injecting melted plastic pellets into the space between two halves of a steel mold. The plastic cools inside the mold and sets into a tub shape.

Cold water cools down the plastic after it has been molded.

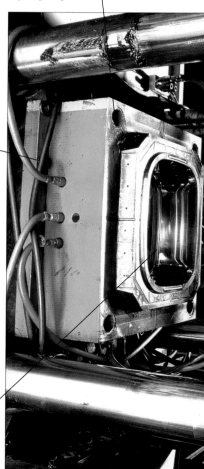

Coloring and plastic pellets are poured in.

Heaters help melt the pellets.

New tub

Mold

A screw pushes the squashed, hot pellets into the mold.

Melted plastic

Each tub needs this amount of coloring.

Two handfuls of polypropylene plastic pellets make each tub.

Temperatures of 535°F (280°C) are needed to melt the pellets and make them flow into this mold.

Don't Throw It Away

You are wasting energy when you throw away plastic bottles or bags. Most plastic "garbage" can be turned into new things, such as the filling for sleeping bags, or fuel bricks, which can burn better than coal!

Plastic Products

Nose cone

A plane's nose cone is made from glass-reinforced plastic, not metal, so that radar waves can pass through it.

Each half of this tub mold weighs as much as 20 eight-year-old children!

Polystyrene is plastic, too.

Many electronic gadgets, such as personal stereos, are made of a tough plastic called ABS.

This half of the mold moves back to let the warm, newly shaped tub drop onto a conveyor belt.

Four tubs can be made in a minute.

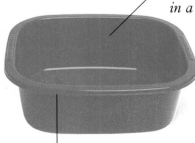

Food slides off the slippery plastic known as Teflon used to coat the surface of nonstick pans.

Under the rim, you may see a line where the two halves of the mold met.

Jets of air blow the tub off the mold when it is finished.

Plastic packaging keeps food fresh longer. For example, polythene bags stop bread from drying out.

Every tub that comes out of this machine is exactly the same shape.

BUILDING

Sears Tower

The invention of new materials and new ways of building has enabled cities to shoot up into the sky. Skyscrapers are not held up by wood, brick, or stone walls, but by strong steel skeletons on which walls and windows are simply hung like curtains. Today's tallest office building, the Sears Tower in Chicago, Illinois, has 110 floors, but engineers now think they could build towers that are six times taller. Imagine taking an elevator to the 660th floor!

Architects designed this skyscraper so that the steel supporting frame is on the outside.

Flexible glue makes sure that the windows don't pop out when the building sways in the wind!

Helicopter pad

47th floor

All large parts, such as wall panels, were brought in by trucks at night. This was the only time that the streets of Hong Kong were empty enough!

Steel was sprayed to keep it from rusting.

Eight steel columns are buried more than 100 feet (30 meters) into solid rock to support the bank.

Growing Up
Building started on the headquarters of the Hong Kong and Shanghai Bank in 1981. It was finished in 1985.

Concrete Creation
Many modern buildings, such as the Sydney Opera House, are made of concrete. This artificial rock is made by roasting clay and limestone to make cement, then adding sand and water. Stretched steel cables inside the concrete stop it from cracking.

Clay

Limestone   Sand

Water

This picture of the bank was taken in February 1984. Steel is being positioned on the 28th floor.

Each of the six red cranes could lift steel as heavy as ten elephants!

The workers clambered about the building on scaffolding made of bamboo.

About 3,500 tons of aluminum were used to make the wall panels.

On some days there were more than 4,500 builders, electricians, painters, and plumbers working on the bank.

Telephone wires, computer links, and electric cables are hidden under the aluminum floor panels.

Fireproof foil wrapping

Building Bridges

Simple beam bridges can span streams, but more complicated bridges are built across wider rivers.

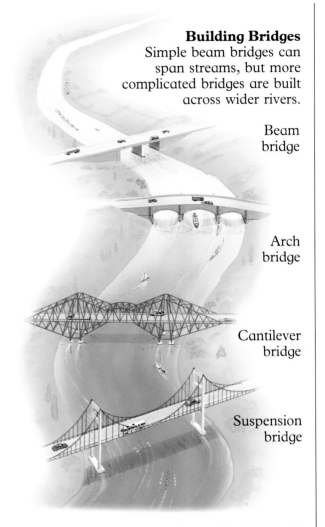

Beam bridge

Arch bridge

Cantilever bridge

Suspension bridge

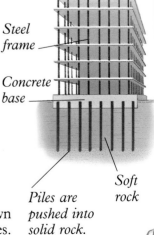

Rock Steady

Buildings need a strong base, or foundation, to stop them from sinking, slipping sideways, or being blown over by the wind. Tall buildings are held down by long steel or cement piles.

Steel frame

Concrete base

Piles are pushed into solid rock.

Soft rock

451

TRANSPORTATION

Transportation is so much a part of our daily lives that most of us take it for granted. Without it, we would all grind to a halt. Millions of people would not get to work, many children would not get to school, and no goods would be delivered to stores. Even the letters we send would never reach their destinations.

Early people relied on animal transportation, and this remained the only way of getting around until about 200 years ago, when the bicycle was invented. Cars didn't come into common use until the first part of the 20th century. Today, though, a jet plane can fly you across the world in hours, and huge spacecraft take astronauts on exploratory journeys to the Moon. Before long, you may be able to book a ticket to Venus or Mars for your vacation!

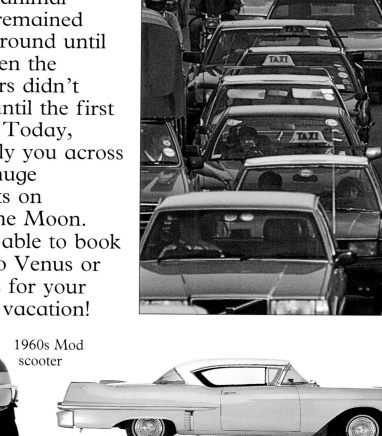

1960s Mod scooter

1950s American car

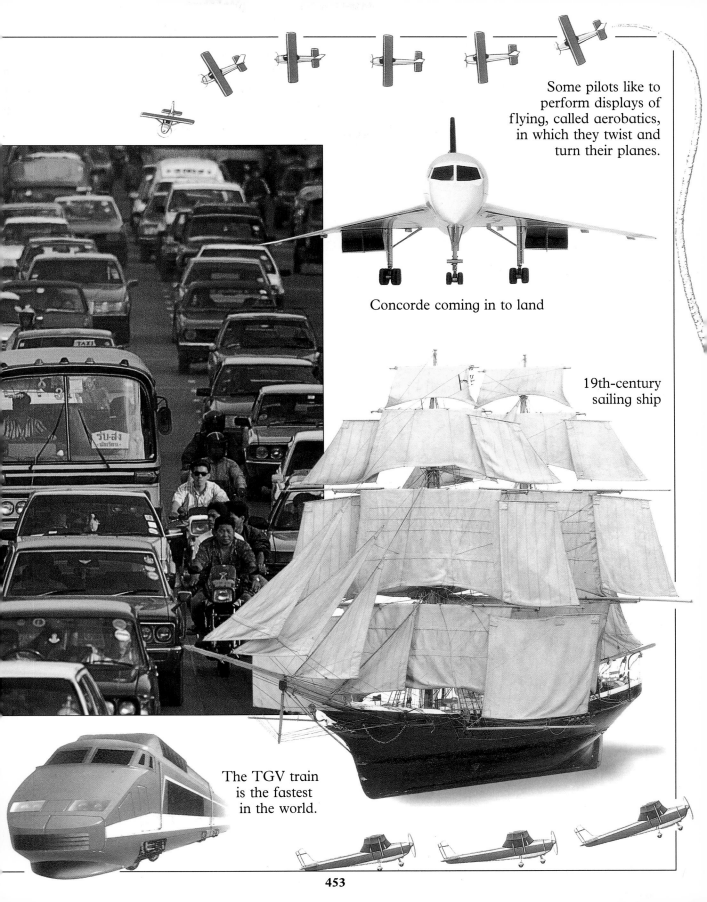

Some pilots like to perform displays of flying, called aerobatics, in which they twist and turn their planes.

Concorde coming in to land

19th-century sailing ship

The TGV train is the fastest in the world.

BIKES

Bicycles were invented in the late 1700s. But they have changed their shape so much over the years that some early bikes, like the "ordinary" bicycle (also called the penny-farthing) look very strange to us today. Power for a bicycle comes from the rider, but sometimes the rider simply runs out of energy! In 1885 a German, Gottlieb Daimler, added an engine to the bicycle and so the motorcycle was invented. The very first motorcycles had tiny steam engines, but today they have gasoline engines and can more than match cars for speed.

Fix It Yourself
Bicycles are simple machines and one of the cheapest types of transportation. Most faults with bicycles are quite simple, too, and can be repaired cheaply and easily by their owners.

Pollution-free Traffic
There are around 1 billion bicycles in China, and in the rush hour in Shanghai there is a crowd of bicycles as people ride home. Imagine the terrible fumes and the effect on the environment if all these cyclists drove cars instead of pollution-free bicycles.

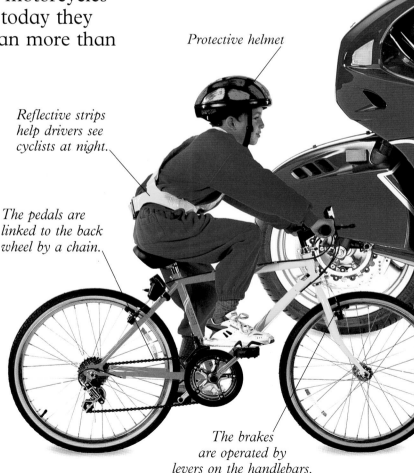

Protective helmet

Reflective strips help drivers see cyclists at night.

The pedals are linked to the back wheel by a chain.

The brakes are operated by levers on the handlebars.

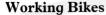

Riding a motorcycle can be dangerous. A rider should wear leather clothes and a special helmet for protection.

Working Bikes

Scooters, which are motorcycles with small engines, are used for many jobs around the world. The Spanish postal system uses yellow scooters for delivering much of its mail. Bicycles can also be adapted for carrying all sorts of things – even pigs!

There are storage compartments under the seat.

This part can be removed to make a seat for a passenger.

Hot Wheels

Motorcycles come in all shapes and sizes for different jobs and sports.

World War II
U.S. Army motorcycle

1966 police motorcycle

1983 Racing motorcycle

The panels of this motorcycle are shaped so the bike will cut smoothly through the air. Its top speed is 148 miles (238 kilometers) per hour.

The motorcycle engine runs on gas.

The exhaust carries waste fumes from the engine.

Motocross motorcycle

CARS TODAY

It is hard to imagine a world without cars. They are all around us and are always being improved to make them more comfortable, more reliable, faster, and safer. And cars must continue to change. There are now so many cars in the world, they are one of the greatest threats to our environment. Experts are constantly thinking of new ways to make affordable cars that use less energy and produce less pollution.

Safety First
Engineers test all new car designs for safety by crashing the car and filming what happens to dummies inside. Some cars have airbags in the steering wheel that inflate in a fraction of a second in a crash.

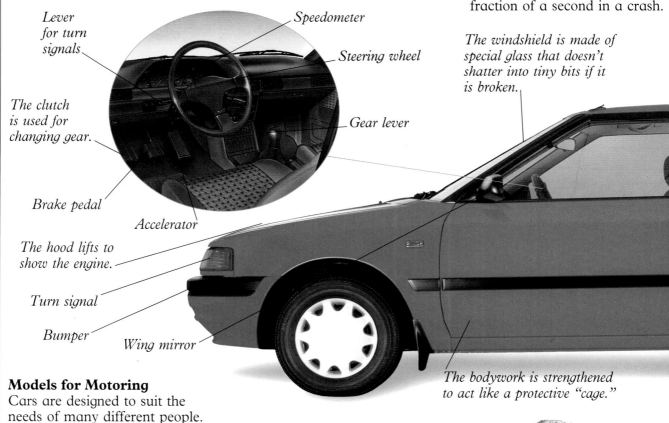

Lever for turn signals

Speedometer

Steering wheel

The clutch is used for changing gear.

Gear lever

Brake pedal

Accelerator

The hood lifts to show the engine.

Turn signal

Bumper

Wing mirror

The windshield is made of special glass that doesn't shatter into tiny bits if it is broken.

The bodywork is strengthened to act like a protective "cage."

Models for Motoring
Cars are designed to suit the needs of many different people.

Compact

Sedan

Sports car

Station wagon

Record Breaker

In 1983, a jet-powered car, *Thrust 2*, reached 633.468 miles (1,019.25 km) per hour. Its aluminum wheels had no tires, as they would rip to shreds at that speed.

Sticking to the Road

Racing cars have low bodies, an airfoil or "wing" at the back that pushes the car down, and wide, gripping tires. These features all help to keep them firmly on the ground at high speeds.

This door opens wide so that luggage can be stored in the back.

Young children must sit in special seats that hold them safely in place.

Future Energy

Most gasoline engines are quite noisy and give off harmful fumes. Quieter and cleaner electric cars are now being designed. But their batteries need recharging after a short distance, so they can only really be used in the city.

Filling up with gas

Knobby tires grip the road in all kinds of weather.

These lights are used to tell the driver behind if the car is going to turn, brake, or reverse. They also make it visible in the dark.

Pickup truck

Stretch limousine

Camper van

TRUCKS

Trucks come in all shapes and sizes and do very different jobs. Some trucks are very specialized and are used for essential services like collecting garbage and fighting fires. Most trucks, however, are used for transporting goods. Trucks are ideal for this job because they can deliver right to the door. Even goods carried by trains, planes, and boats usually need trucks to take them on the last stage of their journey.

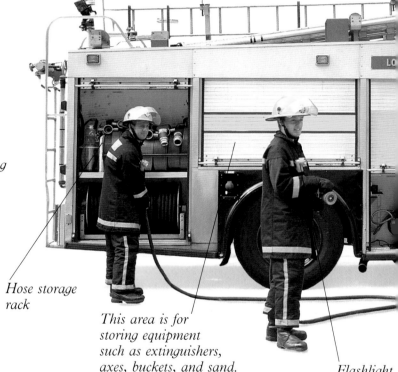

Semitrailer

Tractor

Tight Corners
Long trucks are often made up of two parts that are hinged so they can turn tight corners. The front part is called the tractor, and it pulls a semitrailer.

Floodlight for night emergencies

Warning light

Hose storage rack

This area is for storing equipment such as extinguishers, axes, buckets, and sand.

Flashlight

This gauge shows how much water is left in the fire engine.

Pumps for the fire engine's internal water supply

E852 JYV

Trains without Tracks
Some of the world's biggest trucks are used to transport goods across the desert in Australia. They are called road trains because one truck pulls many trailers.

Big Trucks

Amphibious truck

Cement truck

Dump truck

Car transporter

The equipment for cutting people out of crashed cars is stored here.

The ladder can be raised automatically to a height of 45 feet (13.5 meters).

Air deflector *Container*

A Working Arrangement
The same tractor can hook on semitrailers of many different types, so no journey is wasted. Air deflectors make some loads more streamlined to save fuel.

Tanker

Five crew members can sit in the front and rear of the cab.

This reflective strip makes the fire engine easy to see.

Over the Top
Trucks are not just used for work. People often race them and perform crazy stunts with them, too. This truck, called *Bigfoot*, is demonstrating how to flatten a row of parked cars!

Cab Comforts
At night, long-distance truck drivers usually sleep in their cabs in an area behind the seats. Some have only bunks, but others have televisions, refrigerators, and even stoves.

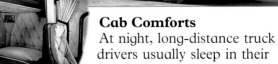

ELECTRIC AND DIESEL TRAINS

In 1964 the Japanese opened the first high-speed electric railway. These "bullet" passenger trains reached speeds of 130 miles (210 km) per hour, a world record at that time. But railway companies earn most of their money by moving goods, called freight. Freight trains are often pulled by diesel locomotives, and they keep a lot of traffic off the roads. The world's longest freight train, in South Africa, had 660 cars and was over 4 miles (7 km) long!

Around the Bend
Most trains only travel at top speeds on straight tracks. But in Italy, a train has been designed that tilts inward on curves, so it doesn't need to slow down much on bends.

Pantographs pick up electricity from overhead wires to power the train.

On electric trains, the section that pulls the cars is called a power car.

The Inter-City Express, or ICE, is an electric train from Germany.

Diesel Power
Like this long Canadian railway convoy, many passengers and goods still travel by cheaper, diesel-powered trains.

The French electric high-speed train is called the TGV. On its regular route, it has a top speed of 185 miles (300 km/h) per hour.

High-speed train routes are expensive to set up because they need a special track that has gentle curves.

At the Controls

The control center for TGVs in Paris is in contact with the driver of every train. Controllers can warn the driver about any problems ahead, such as delays, signal failures, and electrification faults. In this way, the control center keeps the rail network running as smoothly as possible.

Grand Central Station

The largest railway station in the world is Grand Central Station in New York. It has two levels, with 41 tracks on the upper level and 26 tracks on the lower one.

On the Tracks

A great variety of engines and cars use the railway tracks of the world today.

An ICE has two power cars and can carry up to 500 passengers in 12 cars.

Breakdown train

Snowplow

Coast to Coast

One of the world's great railway journeys crosses Australia, from Sydney to Perth. The route covers 2,380 miles (3,968 km) in three days and along the way includes a world record – 297 miles (478 km) of completely straight track!

Passenger train

The streamlined shape of electric trains helps them speed along. In tests, the ICE has reached 214 miles (345 km/h) per hour.

Freight train

UNDERGROUND AND ABOVEGROUND

As a city grows busier, the traffic on the roads becomes heavier and slower. This can often be solved by building a railway across the city – either above or below ground level. There are underground railways, or subways, all over the world, from London, which has the longest with 253 miles (408 km) of route, to Moscow, where stations are like palaces. But they all do the same job – keeping people moving.

No Smoking

The first underground railway opened in London in 1863. It used steam trains, but the smoke often made it impossible to see in the tunnels. The answer was electric trains, which were introduced in 1890.

A Tight Squeeze

So many people travel on the underground railway in Tokyo, Japan, during the rush hour, that special "shovers" are employed to squeeze passengers into the trains.

The London Underground is also called the "tube" because the deep tunnels are built using steel tubing. Tunnels near the surface are dug like ditches and then covered over.

Underground trains are powered by electricity picked up from special rails.

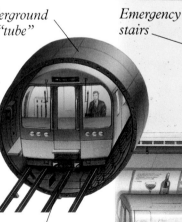

Emergency stairs

Signs tell people where there is an underground station.

London Underground

Riding a Single Rail

It is not always possible to build an expensive underground railway system, so some railways run aboveground. In Sydney, Australia, some trains run on top of a single rail, called a monorail.

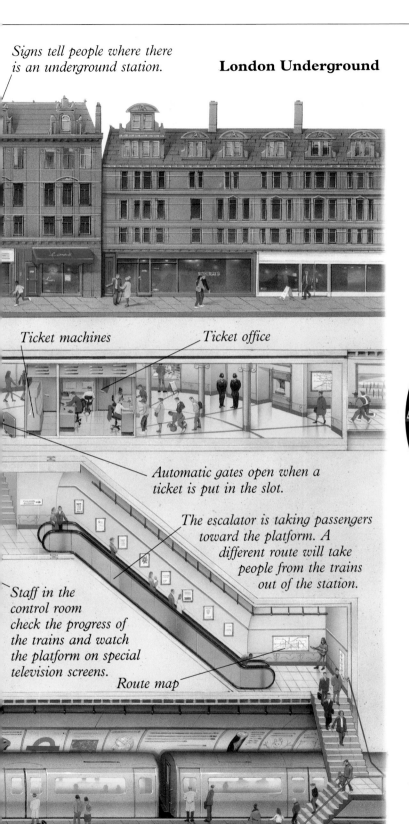

Ticket machines

Ticket office

Automatic gates open when a ticket is put in the slot.

The escalator is taking passengers toward the platform. A different route will take people from the trains out of the station.

Staff in the control room check the progress of the trains and watch the platform on special television screens.

Route map

Hanging On

Not all trains run on top of the rail. The first monorail was built in Wuppertal, Germany, and the electric trains hang from the line.

Going Up!

Special types of transportation are needed for getting people up steep slopes.

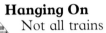

Rack railway

Cable car

Funicular railway

SAILING SHIPS

Nearly three-quarters of the Earth's surface is covered by water, most of it in the seas and oceans. For thousands of years, people have been finding ways to cross this water. At first they built rafts and boats with oars, but around 2900 BC, the Egyptians began to use sails. From then on, sailing ships ruled the seas until a century ago. Today, big ships have engines, but small sailing ships are used for sport, fishing, and local trade.

A Quick Tea
A ship's speed is measured in knots – one knot is about 1.15 miles per hour (1.85 km/h). The fastest sailing ships were clippers, like the *Cutty Sark*, with a top speed of 17 knots or 19 mph (31 km/h). It transported tea from China to England in about 100 days.

China tea bundles

Using a sextant, sailors can find their way when they are in the middle of the ocean from the position of the Sun or the stars.

Sea Charts
The sea often hides dangers, like shallow waters or shipwrecks, so sailors must find their way using sea maps, called charts. These also show other information, like the route a ship must take to avoid other ships.

Tall Ships
Many of the great ships of the past have been restored and are used today for special "Tall Ships" races.

Across the Ocean
For hundreds of years, sailing ships have traveled the oceans of the world for exploration, trade, and war.

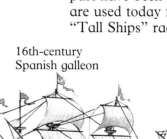

15th-century Portuguese caravel

16th-century Spanish galleon

17th-century merchant ship

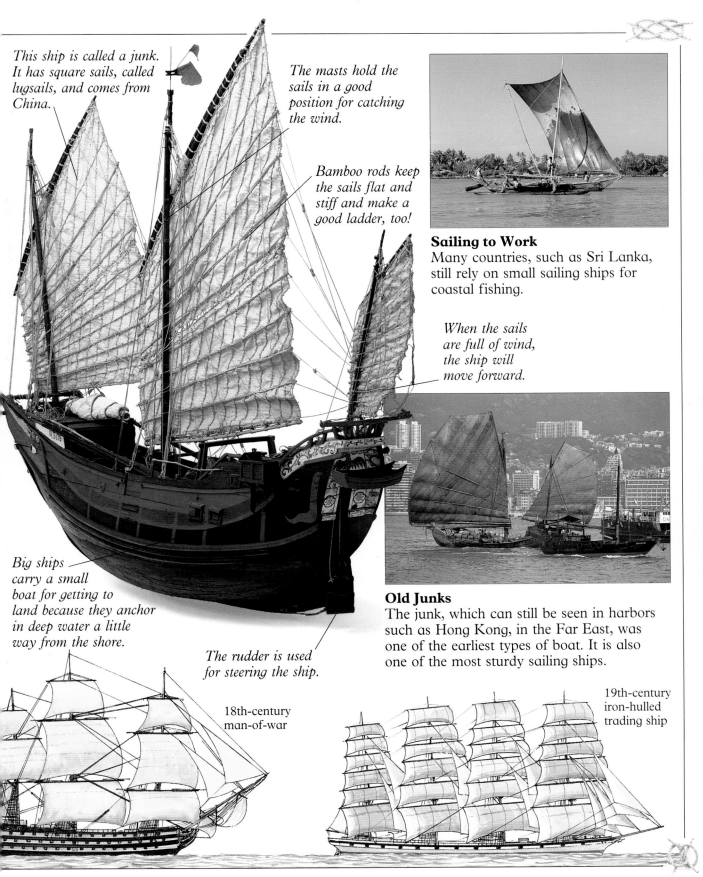

This ship is called a junk. It has square sails, called lugsails, and comes from China.

The masts hold the sails in a good position for catching the wind.

Bamboo rods keep the sails flat and stiff and make a good ladder, too!

Sailing to Work
Many countries, such as Sri Lanka, still rely on small sailing ships for coastal fishing.

When the sails are full of wind, the ship will move forward.

Big ships carry a small boat for getting to land because they anchor in deep water a little way from the shore.

The rudder is used for steering the ship.

18th-century man-of-war

Old Junks
The junk, which can still be seen in harbors such as Hong Kong, in the Far East, was one of the earliest types of boat. It is also one of the most sturdy sailing ships.

19th-century iron-hulled trading ship

SHIPS WITH ENGINES

Wind is not a very reliable form of power – sometimes it blows from the wrong direction, and sometimes it does not blow at all! But from around 1800, steam engines were used to turn paddle wheels or propellers. Steam power moved ships faster and was a more reliable way of transporting people and goods. Today, ships use mainly diesel engines, and their most important job is carrying cargo.

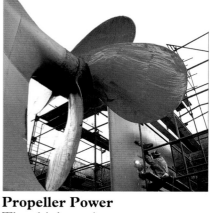

Propeller Power
The ship's engine turns a propeller at the back of the ship. This pushes the ship forward. On a big ship, these propellers can be enormous.

This part of the ship is called the bridge.

The ship is steered from the wheelhouse.

Living quarters

The cruising speed of a ship like this is about 15 knots or 17 miles per hour (28 km/h).

In the Dock
In modern ports, like Singapore, most cargo arrives on trucks and trains and is already packed in containers. These are stored on the dockside and can then be neatly loaded onto the ships using cranes.

Pull and Tug
Big ships are difficult to control and from full speed can take several miles just to stop. So in ports and harbors small, powerful boats, called tugs, help push and pull big ships safely into position.

A Vacation at Sea

Cruise ships offer passengers a luxury vacation as they travel. One of the best-known cruisers today is the *Queen Elizabeth II*. It can hold 1,800 passengers and 1,100 crew.

A Loading Line

On a ship's side is a row of lines, called a Plimsoll mark. One or another particular line must always be above water, or the ship may sink. There are several lines because ships float at different levels in salt and fresh water, in summer and winter, and in the tropics and the North Atlantic.

Plimsoll mark

The front of the ship is marked with a sighting mast. Otherwise, it would be hidden from the crew in the bridge by the cargo on the deck.

Containers are stacked in racks on the deck.

GRACECHURCH CROWN

Save Our Souls

In an emergency at sea, passengers and crew put on life jackets and send an SOS signal. Rescue is carried out by a lifeboat crew.

Ship Shapes

The seaways of the world are busy with ships of all shapes and sizes.

Naval frigate

Roll-on/roll-off ferry

Oil tanker

SKIMMING OVER THE WATER

Most water travel is not very fast because the water itself slows a boat down. But Hovercraft and hydrofoils just skim across the water, so they can travel at great speed. A Hovercraft is not really a boat because it hovers above the surface of the water. It is also amphibious, which means that it can travel on both land and water.

The car deck will hold up to 60 cars.

This Hovercraft can travel at a speed of 60 knots (69 miles or 111 km per hour) in calm waters.

Water Rafting
People get around the reed beds and watery forests of the Everglades in Florida using flat-bottomed rafts. These have raised fan motors that do not get stuck in the weeds.

Propellers push the vehicle forward.

What a Drag!
Water-skiers can speed over the water, but will slow right down if they fall in. This is because water is over 800 times denser than air.

Control cabin

The cars drive out of this door.

Up and Away

Hovercraft are also known as air-cushion vehicles because they float on a cushion of air.

When the Hovercraft is sitting still on the tarmac, the skirt is flat and empty.

The engines start, and the skirt fills with air to become a thick cushion.

Hovering just above the surface of the water, the vehicle speeds on its way.

The cars drive in through the doors at the back.

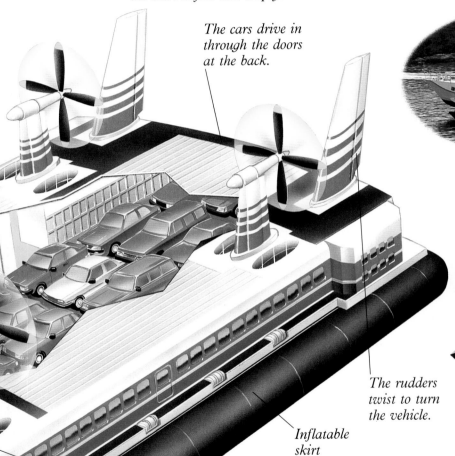

Thin Fins

Fins under a hydrofoil lift it out of the water so that it can skim over the surface at great speed.

The rudders twist to turn the vehicle.

Inflatable skirt

Riding High

Catamarans have two thin hulls, so there is very little of the boat in the water to slow it down.

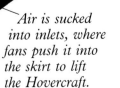

Air is sucked into inlets, where fans push it into the skirt to lift the Hovercraft.

A Hovercraft like this can carry over 400 passengers.

Super Speedy

Superboats have very powerful engines. They race at speeds of around 155 miles (250 km) an hour – so fast that their hulls rise out of the water.

PLANES WITH PROPELLERS

For hundreds of years, people tried in all sorts of crazy ways to fly like birds and insects. But it was not until 1903, when the Wright brothers attached a propeller to a small gasoline engine, that people first managed to control a plane's takeoff and landing. Today, huge jet planes can fly hundreds of people around the world, but smaller, cheaper planes with propellers are still the best form of transportation for many jobs.

These propellers have three blades. As they spin around, they move the plane forward.

Skis have been attached to the wheels so the plane can land and take off from the snow.

Fresh Air!
In 1927 Charles Lindbergh made the first nonstop solo flight across the Atlantic in the *Spirit of St. Louis*. Like all pilots at that time, Lindbergh had to wear warm leather flying gear such as a helmet, gloves, a coat, and boots, to protect himself from the cold.

Open cockpit

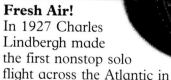

To the Rescue
Light aircraft are relatively cheap to run and are ideal for transportation in large remote areas. In parts of Africa, for example, the only way doctors can get to their patients quickly is by plane.

Short Takeoff
Small planes only need short runways. This means they can go to places where larger planes could not land, like this grassy airstrip on a mountainside in Nepal.

War and Peace
Over the years, there have been many types of propeller-driven aircraft.

World War I triplane

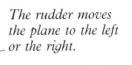

This plane can seat up to 19 passengers.

The rudder moves the plane to the left or the right.

World War II Spitfire

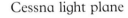

This tail plane keeps the plane stable. Hinged flaps at the back of it move up and down to make the plane climb or dive.

Cessna light plane

This twin-engined "Otter" can take off and land in a small space.

Seaplane

Fighting Flames
This Canadair plane is specially designed for fighting forest fires.

Flying slowly and quite low, the pilot opens a hatch to release the water.

The pilot prepares to lower two pipes to scoop up water into the body of the plane.

As the plane skims the surface, up to 1,600 gallons (6,400 liters) of water are forced into the tanks in ten seconds.

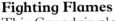

PASSENGER PLANES

The biggest airliner today is the Boeing 747, which can carry around 500 passengers. Because of its size and powerful jet engines, it is called the jumbo jet. Big planes have made flying much cheaper, and millions of people pass through the world's airports each year. The passengers simply step onto a waiting plane, but many jobs must be done to ensure every flight takes off and lands safely.

Join the Line
The world's busiest airport is Chicago's O'Hare International Airport in Illinois. Around 110 aircraft arrive or depart each hour. As at many busy airports, planes have to line up at a "holding point" before they can take off.

A Job for Batman
Once a plane has landed, a marshal helps to park it by signaling instructions to the pilot with brightly colored bats.

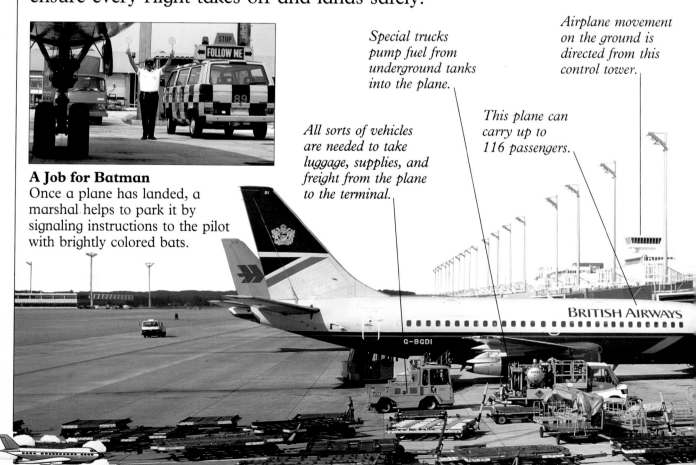

Special trucks pump fuel from underground tanks into the plane.

Airplane movement on the ground is directed from this control tower.

All sorts of vehicles are needed to take luggage, supplies, and freight from the plane to the terminal.

This plane can carry up to 116 passengers.

Supersonic

The fastest passenger plane today is the Concorde, which has a cruising speed of 1,335 miles (2,150 km) per hour. Because this is faster than the speed of sound, it is called a supersonic plane.

A Room with a View

At peak hours in major international airports, planes are landing almost every 60 seconds, while other planes wait to take off. The takeoff and landing of these planes is carefully supervised from the control tower.

Curious Cargo

Goods may be loaded onto a plane with a scissor-action crane or a conveyor belt. Some cargo simply walks in and out of the hold!

A Good Checkup

Before each journey, the ground crews give the plane a good cleaning inside and sometimes outside! Engineers check that the plane is in perfect working order.

A loading tunnel links the plane to the terminal.

From this control tower, staff direct takeoffs and landings.

Service vehicle

Passengers check in for their flights in the airport terminal.

INDEX

A

abacus 293
Aboriginals 301, 390, 391
ABS plastic 449
acacia 63, 79
Achilles tendon 241
Africa 15, 325, 348, 360-63
    animals of 192, 193, 194,
       199, 208, 210
    birds of 173, 174, 182
Afrikaners 289, 363
afterlife 258
agave 62
air 402-403, 412
    pollution 357, 456
aircraft carrier 422-23
air-cushion vehicles 468-69
airports 472-73
Alaska, U.S.A. 351
albatross, Royal 158
Albertosaurus 149
Aldrin, Edward "Buzz" 40
algae 57
Algeria 360, 361
allium 73
Allosaurus 149
alloys 442
alpaca 213
alpine plants 71
aluminum 443, 451
American civil war 290-91
Amerindians 251, 276-77,
    352-53, 356, 359
amphibians 57, 151, 397
Amsterdam, Holland 370
Amundsen, Roald 351
Andes mountains, South
    America 273, 359
anglerfish 85
animals 57, 87, 259, 396
    breathing of 232-33, 402,
      403
    See also individual species
Antarctica 25, 178-79, 350-51
antelopes 187, 214
anther 72, 74
anthracite coal 431
ants 108, 120-21
apartheid 363
Apatosaurus 149
apes 198-99
Apollo space program 36-39,
    40-41
apple 60
    arachnids 108
    Arctic 25, 58, 160-61,
      350-51, 366-67
    Arizona, U.S.A. 51,
      151
    Armstrong, Neil 40

Arnolfini Marriage, The (J. van
    Eyck) 310-11
artists 271, 310-11,
    383, 390
    Cook's voyages 282-83
Ashanti kingdom
    278-79
Asia 325, 326, 380-81
    animals of 192, 193, 194
    See also individual
      countries
ass 208-209
Astaire, Fred 305
asteroids 35, 50, 51, 150-51
astrolabe 267
astronomy 52-53
athletics 260-61
atoms 398, 436-37
Australia 11, 79, 288, 390-91
    animals of 175, 188-89
    food production in 336, 346
Austria 376-77
automatons 426
avocado 77, 273
Ayers Rock, Australia 11

B

baby development 246-47
bacteria 57, 396
badger 174, 397
bagpipes 301
Bakelite 448
baleen whales 102-103
Bali, dancers 389
ballet 295, 306-307
bamboo 69, 451
Bangkok, Thailand 389
Bangladesh 31, 386, 387
banjo 301
banyan 68
baobab 69
barley 322-23
Barosaurus 133, 140, 142-43
bats 75, 154, 187, 202-203, 413
batteries 405, 417
Bavarisaurus 146
bazaar, Istanbul 364
beans 273, 343
bears 196-97
Beauchamps (ballet teacher)
    306
beaver 206, 207
bedouin Arabs 364-65
bee-eaters 157, 175
beef 330-31
bees 74, 108, 124-25
beetles 109, 112-13
beetroot 340
Belgium 370
Berbers 360
Bering Strait 350
Bhutan 386
Bhutto, Benazir 387
Bible, Gutenberg 270
bicycles 418-19, 454-55

Billund, Denmark 367
biology 394, 396
birch 63, 82
bird of paradise 173
bird-of-paradise flower 75
birds 57, 86, 151, 152-85, 221,
    225, 397
birds of prey 160, 180-81
birth 247
bison 214-15, 253
bitumen 447
bivalves 88-89
blackberry 58
blackbuck 215
black holes 55
blood 224, 230-31
blood gun 319
blue tits 152, 159, 164
blue whale 84
Boeing 747 aircraft 472
boiling points 401
Bolivia 358, 359
bombs, nuclear 422-23
bones 242, 243, 244-45
books 267, 270
botany 396
bovids 214-15
Brachiosaurus 140-41, 142
Brahmaputra River, India 31
braille 226
brain 224-25, 226, 240
brass 442
Brazil 201, 346, 358, 359
breathing 96, 232-33, 396,
    402-403
bridges 451
British Isles 286-87, 368-69
Bronzino, Il 310
bubble shells 88
Buddhism 274, 383, 389
budgerigar 152
buffalo, African 215
buildings 450-51
Bunyols festival, Spain 372
bushmaster snake 346
Bushmen, Kalahari 362
bustard, Kori 175
butane 434
butterflies 74, 109, 114-17
butterfly fish 283
butterwort 67

C

cable car 463
cactus 58, 61, 69
calcite 22, 23
California, U.S.A. 435
camels 212-13, 268-69, 330,
    364
Camarasaurus 132
cameras 314-15
Canada 288, 346, 352-53
    animals 194, 197, 215
Canadair plane 471
"canaphone" 220

capsicum 342
capybara 206
caravans 269
carbohydrates 238, 239
Caribbean 356-57
carnivores 194-95, 196
carnivorous plants 66-67
carnosaurs 149
carrot 340
cars 355, 444-45, 456-57
cartilage 97, 100, 245
car transporter 459
cassava 341
cassowary 183
castles 266, 376, 379
cat cracker 447
Cat on a Hot Tin Roof
    (play) 298
Çatal Hüyük, Turkey 255
catamarans 469
caterpillars 109, 114-15, 116
Catherine II, Empress of
    Russia 280-81
catkins 74
cats 192-93, 225, 259, 263, 413
cattle 212, 214, 254, 330-31,
    358
caves 22-23, 26
    paintings 252-53, 310
cavy, rock 206
Central America 272-73, 325,
    356-57
cephalopods 95
Ceratosaurus 135, 149
cereals 254, 256, 322-27
Cessna light plane 471
Chambers, Tommy 418
Chaplin, Charlie 316
cheetah 192-93
chemical industries 432, 446-49
chemistry 394
Chicago O'Hare airport 472
chilies 273, 343
chimpanzee 198-99
China 36, 310, 382-83, 422,
    454, 465
    Marco Polo and 268-69
    plays and 299
    sugar production in 346
chinchilla 206
chipmunk 206
chiton 89
chlorine 447
chlorophyll 64-65
chocolate 272, 344, 370
Christians 266, 365
Christ statue, Rio de Janeiro,
    Brazil 358
chromium 443
chrysalis 116
circulatory system 219, 230-31
citrus fruits 336-37
clam 89
clementine 337
cliffs 23, 26-27, 171

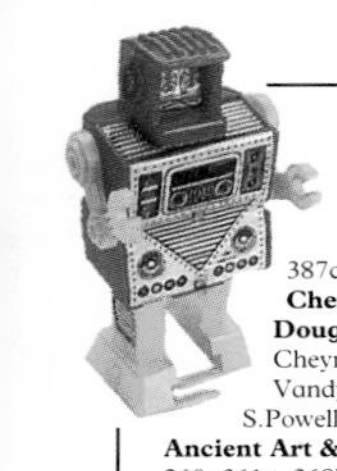

PICTURE CREDITS

A.F.P. Photo, Paris: Mufti Munir 387cra; **Air France:** 473tl; **Bryan & Cherry Alexander:** 366clb, 367tl; **Doug Allan:** 179cr, c; **Allsport:** Russell Cheyne 391cr, Yann Guichaoua/Agence Vandystadt 404tl, D.Klutho 459bl, S.Powell 261br, Pascal Rondeau 415br; **Ancient Art & Architecture Collection:** 256tl, 260, 261c, 268bl; **Aquila Photographics:** C.Greaves 167tl; **Archiv Fur Kunst und Geschichte, Berlin:** Musee du Louvre, Paris 310bcr; **Ardea:** G.K.Brown 177l, D.Parer & E.Parer-Cook 13t, K.W.Fink 209cra, F.Gohier 102b, 103t, 105tc, 213cb, C.Haagner 13b, J.M.Labat 188c, Mike Osmond/Auscape Int.102-3, J.Swedberg 170c, R.&V.Taylor 98b, A.Warren 203cr, 199b, A.Wearing 239tr; **Art Directors:** 352c, 375tc; **ASAP:** Y.Mazur 365tl; **Australia House:** 461bc; **Australian Picture Library:** 371r; **Aviation Picture Library:** Austin J.Brown 449tr, 472tr, 473tc.

Hans Banziger: 119cl; Barnaby's Picture Library: 349br; **Michael Benton:** 137clb; **BFI Stills, Posters & Designs:** courtesy Productions La Fete Inc., Montreal, Canada 378tr; **Bibliotheque Nationale, Paris:** 267c; **Biofotos:** Heather Angel 29cr, 62br, 68br, 71tl, 78c, 78clb, 81tc, 106tr, 344cb, Bryn Campbell 25, Brian Rogers 29cra; **Birmingham Museum:** 349crb; **Black Hills Institute of Geological Research:** Ed Gerken 138bl, 139tc; **BNFL:** 437crb; **Bridgeman Art Library:** Bibliotheque National, Paris 310cl, Christies, London © 1993. Pollock-Krasner Foundation 311cr, Forbes Magazine Collection 287tc, Galleria del'Academia, Florence 313l, Galleria degli Uffizi, Florence 310bcl, Giraudon 306tl, 310tl, 310bl, Giraudon/Louvre 303br, National Palace, Mexico City 272br, Guildhall Library 462c, Musee Rodin, Paris 295br, National Gallery, London 311tl, Phillips 311bc, Private Collection © ADAGP, Paris & DACS, London 1993 311bl, Private Collection © DACS 1993 310br, Tretyakov Gallery 280bl; **British Aerospace:** 424bl; **British Coal:** 430tl, cr, 431tc, cl, bl; **British Gas:** 434br, 435c; **Courtesy Trustees of the British Museum:** 250bl, 251bl, br, 258tl, 258-9c&b, 259r, 260t, 260-1t, 278-9t, 357cra; **British Steel:** 430cl, 442br, 443bl; **John Brown:** 447bra.

J.Allan Cash: 433c; **Christie's Colour Library:** 293bc; **Bruce Coleman Ltd.:** 170-1, D.Austen 212c, Jan & Des Bartlett 176cl, 182b, Erwin & Peggy Bauer 141tc, 175cr, M.N.Boulton 81tr, Mr.J.Brackenbury 122t, J.Burton 67bl, 78cla, 113bl, 194-5t, 204cr, J.Cancalosi 325cl, G.Cappelli 339cr, D.Chouston 180tr, A.Compost 198-9t, E.Crichton 114b, 145bl, 402cl, G.Cubitt 66tr, 69cr, P.Davey 216cl, A.Davies 200c, F.Erize 107tl, 178tr, 206tr, Dr.I.Everson 102t, MPL Fogden 75tl, 118tr, Jeff Foott 61cl, 104-5b, 176r, 209ca, C.B. & D.W. Frith 64cl, F.Furlong 170tl, J.Grayson 109tr, D.Green 158tr, F.Greenaway 171tc, U.Hirsch 72br, M.P.Kahl 65tr, S.C.Kaufman 67br, 201br, 347bc, S.J.Krasemann 115cb, 202-3b, 331bc, F.Labhardt 126cb, H.Lange 61tr, G.Langsbury 171cl, F.Lanting 116br, L.Lee Rue 66br, 329tr, 363bl, M.Timothy O'Keefe 70bl, 336bc, W.S.Paton 193tr, M.R.Phicton 137cra, 210cr, D.& M.Plage 10b, 217cr, Dr.E.Pott 74cr, 329br, Dr.Sandro Prato 66-7c, 115cr, M.P.Price 58-9c, A.J.Purcell 74bl, 109tc, M.Read 130t, H.Reinhard 60bl, 71tr, 106bl, 210tr, 215tc, F.Sauer 123b, J.Shaw 162-3, 151tl, K.Taylor 117bl, 123br, 127tl, 258bl, N.Tomalin 192cr, M.Viard 335crb, J.Visser 205cr, R.William 119tl, 141c, K.Wothe 69bc, 201tl; **Bruce Coleman Inc.:** 113c; **Colorsport:** 375crb, Sipa Sport 469cr; **Comstock Inc.:** George Lepp 159cl.

James Davis Worldwide: 388cla; **Derngate Theatre, Northampton** 380tr; **DLP:** D.Heald 342-3c, D.Phillips 301tc.

Ebenezer Pictures: J.Browne 383crb; **Ecoscene:** 449tc; **E.T.Archive:** 288bc, 464tr, National Maritime Museum 282tl, bl, V.&A.Museum 251cr; **Mary Evans Picture Library:** 284l, 285tr&b, 287b, 314tr, 357cl, 376clb, 422clb.

Gary Farr: Photo appears courtesy of New Line Cinema Corp. 317crb; **Chris Fairclough:** 313tr, 330bl, 447tc; **Ffotograff:** Patricia Athie 369crb; **Fine Art Photographic:** 286bl; **Focus, Argentina:** 358-9c; **Michael & Patricia Fogden:** 77br; **Foods from Spain:** 337bl; **Ford Motor Co.Ltd.:** 456tr; **Werner Forman Archive:** Asantehene of Kumase 279crb, Statens Historiska Museum, Stockholm 264tl, Plains Indian Museum, Buffalo Bill Historical Centre, Wyoming 276; **French Railways Ltd.:** 461tr; **Fullwood:** 331c.

Christina Gaiscoigne: 265tr; **John Paul Getty Museum:** 262cr; **David Gillette, PhD:** 137tl; **Giraudon:** 285tc; **Ronald Grant Archive:** 316bc, 317br; EMI Film Productions Ltd.: 306c, © Saul Zaentz Co., all rights reserved, Francois Duhamel: 318bl; **S.&R.Greenhill:** 220cl, 236cl; **Geoscience Features:** 16c, 17, 23tr, 151cl.

Rafn Hafnfjord: 15; **Robert Harding Picture Library:** 23tl, 27b, 248-9, 252tr, 256br, 266bl, 271cl, 282-3b, 316tr, 333tc, 342tl, br, 347tr, 353cl, 362cla, 366tr, 372cla, 374tr, 386clb, 387crb, 434cl, 439cr, 444cl, bl, 445tc, 451br, 455tc, 463cr, 496tl, tc, tr, Bildagentur Schuster/Meier 379bl, M.J. Bramwell 470-1c, P.Craven 380-1, G.Heller 385bc, J.Green 471tl, D.Maxwell 232cl, C.Rennie 328bl, 381c, 455tr, W.Rawlings 370cb, V.Southwell 257tr, G.M.Wilkins 469cra, A.Wolf 460bc, A.Woolfitt 370-1bc, 386cla; **Michael Holford:** 50c, 250-1c, 253tc, 261cr, 274bl, 287cra, crb, 375bl; **Holt Studios International:** R.Anthony 325bl, D.Donne Bryant 327tr, cr, Nigel Cattlin 68cl, 78cr, bc, 323cb, 326-7t, 333cla, 334bl, 340br, 343cla, 344c, 345br, Jurgen Dielenschneider 329bc, Primrose Peacock 341c, Inga Spence 320-1c, 338-9tc, 339clb; **House of Marbles:** 292tl,tr; **Dianne R.Hughes:** Biological Sciences, Macquarie University, NSW Australia 99tc; **Hulton-Deutsch Collection Ltd.:** 304c, 470bl; **Robert Hunt Library:** 423br; **Hunterian Museum, Glasgow:** Malcolm McCleod 278b; **Hutchison Picture Library:** 296clb, 361tl, Sarah Errington 363tl, V.Lamont 325c, L.McIntyre 325tl, M.MacIntyre 302bl, Stephen Pern 381cl, John Ryle 358cl.

The Image Bank: 444tr, S.Allen 457crb, D.Berwin 366c, I.Block 390cl, J.Cartier 407tr, A.Caulfield 450tl, 371cl, G.M.Corian 137cla, A.Choisnet 308cr, M.Coyne 450cl, G.V.Faint 280-1t, 460bla, D.Fisher 412tr, Fotoworld 377cl, Di Giacomo 332-3, T.King 468bl, R.Lockyer 308tl, N.Mascardi 392-3, R.Phillips 301tr, Andrea Pistolesi 356b, Barrie Rokeach 354br, Marc Romanelli 301c, Guido Alberto Rossi 212tr, 288tl, Schloter 445br, H.Schoenbeck 460-1bc, M.Skaryd 472cl, J.Smith 443tr, Harald Sund 288-9, 383cb, Dag Sundberg 367c, Jack Ward 209br, Frank Wing 338c; **Images:** 2tr, 229cl, 234cla, 244cr, 287cr, 370cl, 390cla, 396tc, 404bl, 466tr, 467tr, cr, 472tl; **Impact:** Sergio Dorantes 357tl, M.McQueen 422cla, G.Mendel 363br, M.Mirecki 373tc, B.Rybolt 298tr.

Jacana: A.Le Garsmeur 382tr, Jean-Michel Labat 157tc, Jean-Philippe Varin 159t.

Kobal Collection: 305tr, 316cl, 317cra, tr, Warner Bros. 318br.

Ian Lambot: 451tl,cl,bl; **Frank Lane Picture Agency:** E.&D.Hosking 191tl, S.McCutcheon 28clb, M.Newman 160-1, R.Van Nostrand 124tcr, Fritz Polking 174bl, Len Robinson 185t; **La Vie Du Rail:** 460tr; **Legoland:** 367cl; **Courtesy of Lucasfilms Ltd.:** 'Jedi' ™&©. Lucasfilm Ltd.(LFL)1983 all rights reserved 319cl.

Magnum: Abbas 360tr, 387c, Eve Arnold 382-3c, Bruno Barbey 294-5c, 373c, Fred Mayer 373bl, James Nachtwey 357cr, Chris Steele Perkins 359c, Raghu Rai 3c, 386-7c; **MARS:** U.S.Navy 422tr; **Marshall Cavendish Picture Library:** 267tr, 292tc, bc, Palazzo Tursi di Genoa, Servizio Beni Culturali 268tl; **Michelin:** 415cr; **Massey-Ferguson Ltd.:** 321br; **Michel Muller:** © Henry Moore Foundation (1993) reproduction by kind permission of the Henry Moore Foundation 312br; **Museum of Automata:** York © 426cl; **Museum of London:** 250br, 262b, 263; **Museum of the Rockies:** Bruce Selyem 136b, 137br.

NASA: 37l, 39br, 40, 42br, 43t, 45t, 46tr, 47tl, br, 49c, 51br, 53cr, 150cr, 415tr; **©National Geographic Society:** Wilbur E.Garrett 350cl; **National Maritime Museum:** 453br; **National Motor Museum:** 457tl; **National Palace Collection:** Museum of Taipei, Taiwan, Republic of China 269br; **Natural History Museum, London:** 137crb, cb, 138cl, bc, br, 283r; **Nature Photographers Ltd.:** Frank Blackburn 180bl, K.Carlson 166bl, Hugh Clark 163b, A.Cleare 171c, E.Janes 167b, P.Sterry 335br; **Peter Newark's American Pictures:** 285crb, 291tr; **Network Photographers:** Nikolai Ignatiev 350tr, 380cl, J.Leighton 376cla, Barry Lewis 378b, 379cr, Paul Lowe 364cla, Dod Miller 369tl, Laurie Sparham 379cr; **NHPA:** Agence Nature 214c, H.Ausloos 210clb, 403cr, A.Bannister 75tr, 114tl, 130bl, G.Bernard 71br, Bishop 60cl, N.A.Callow 68tl, L.Campbell 69crb, S.Dalton 75b, 110tl, tr, 116tr, 123bl, 190clb, tr, 204cla, c, M.Danegger 152-3, J.B.Free 69tc, S.Krasemann 113tr, M.Leech 205l, D.Middleton 83r, M.Morecombe 74tl, L.H.Newman 175cl, A.Papaziar 101b, S.Robinson 198c, John Shaw 74c, 82cl, 161c, 196cr, R.Tidman 159cr, M.Tweedie 113crb, D.Watts 178cr, Martin Wendler 122c, 183cl; **Nissan:** 445tr.

Oxford Scientific Films: D.Allan 107c, 179tr, Animals, Animals 113cl, Breck P.Kent 214-5b, K.Atkinson 88c, S.Bebb 20l, Hans & Judy Beste 202bl, G.Bernard 172tr, Neil Bromhall/Genesis Film 247t, S.Camazine 128tr, J.C.Cannon 179tl, Densey Clyne 121c, M.Colbeck 33c, D.Dale Photo Researchers Inc. 110b, B.Fredrick 126c, MPL Fogden 121t, 346tr, D.Lee 412bl, S.Littlewood 114cl, R.Lynn/Photo Researchers 148tl, J.&C. McDonald 188-9c, T.McHugh 18bl, 147tr, G.A.McLean 124tr, Mantis Wildlife Films 114tr, 120tl, T.Middleton 23b, C.Milkins 128cl, S.Osolinski 192-3t, P.Parks 86clb, c, 86-7b, t, H.Reinhard 333c, J.H.Robinson 156bl, F.Schneidermeyer 104cl, David Thompson 73tr, 125bl, R.Toms 177r.

Panos: 363tr, Wang Gang Feng 454bl; **Peabody Museum, Yale University:** 133bl, J.H.Ostrom 146br; **The Performing Arts Library:** Clive Barda 309br; **Dr.Chris Pellant:** 26; **Photostage:** Donald Cooper 298bla, bl, 299br, 309tl, c, bl; **Pictor:** 31t; **Picturepoint Ltd.:** 361ca; **Pitt Rivers Museum, University of Oxford:** 251tl; **Planet Earth Pictures:** K.Ammann 3br, 210-1b, G.Bell 188-9cb, J.Braagirde 207cr, J.Brandenburg 194cl, 196tr, M.Clay 192cl, L.Collier 93bc, R.Coomber 166tl, 217tc, P.David 85cr, Georgette Douwma 145cl, H.C.Heap 32, A.Kerstich 94cl, K.Lucas 95tc, br, 108-9, 198tl, J.Lythgoe 20r, R.Matthews 201c, D.Perrine 88-9b, Christian Petron 89br, M.Potts 31b, R.S.Rogoff 195tr, A.&M.Shah 192bc, P.Scoones 84-5c, J.Scott 216cb, 216-7c, P.Stephenson 214bl, H.Voigtmann 100b, J.D.Watt 104bl, Norbert Wu 94tr; **Popperfoto:** 425tc, 470cl; **Premaphotos:** K.G.Preston-Mafham 119b, 120bl, 122br, 131t, R.A.Preston-Mafham 128tl; **Quadrant:** 473cr, A.Dalton 446-7c.

©John Reader 1994: 359cra, 360clb, 362cr, b, 364bc, 364-5c, 367tc, 373tl, 386cb, c, 389tr, 391c; **Renault:** 457cr; **Retna Pictures:** J.Welsby 375c; **Rex Features Ltd.:** M.Friedel 374cl, Stevens/Zihnioglu/ Sipa Press 375cla, Stills/Pat/Arnal 370cr; **Ludwig Richter:** Guttenberg Museum 270c, bl, br; **Royal Botanic Gardens, Kew:** David Cutler 70br; **Royal Collection, St.James' Palace ©Her Majesty The Queen:** 280tl; **Royal Geographical Society:** 351tc; **Royal Museums of Scotland:** 250bcr.

Scala: 253tr, Iraq Museum, Baghdad 257tl, Museo della Scienza, Firenze 271tr, Museo Vinciano, Vinci 271cb; **Science Museum:** 36c, 39bl, 42cr, 44-45; **Science Photo Library:** 53tr, A.Bartel 422bl, 438cl, 440-1tc, Dr.Jeremy Burgess 399cr, CNRI 243tr, 394-5c, T.Craddock 430tr, 438-9bc, Dr.Martin N.England 55br, Fred Espanak 34-5c, 55cr, European Space Agency 8b, 10tl, Dr.G.Feldman/ NASA/GSFC 87tr, S.Fraser 1c, 447tr, D.A.Hardy 55c, 150bc, Adam Hart-Davies 410cr, Kapteyn Laboratorium 52b, Dr.J.Lorre 55cra, M.Marten 437tl, J.Mason 437tr, A.McClenaghan 411br, P.Menzel 424bc, 460tlb, Prof.Motta/Dept. of Anatomy, University 'La Sapienza', Rome 235br, 238tl, 410tr, Prof.E. Mueller 151c, NASA 10-1c, 37r, 49t, 50b, 151b, 412tl, 414, 442bl, NOAO 54b, Claude Nuridsany & Marie Perennou 28tl, cla, bl, D.Parker 18, Max Planck Institute 52t, E.Pritchard 452-3c, R.Ressmeyer, Starlight 436bc, 437bc, R.Royer 54-5, J.Sanford 51t, c, 53tl, Dr.R.Schild/ Smithsonian Astrophysical Observatory 54t, Dr.R.Spicer 150cl, S.Terry 21r, 42bc, A.Tsiaras 413tr, U.S.Dept.of Energy 427br, 436tl, 437cr, U.S.Naval Observatory 53br, Tom Van Sant/Geosphere, Santa Monica 4tl, 11crb, bc, D.Vaughan 351c, V.Vick 351cr, E.Viktor 150bl; **Schwangau/Ostallgau, Germany:** Tanner Nesselwang 376bl; **Sea Containers:** 469crb; **Survival Anglia:** Jeff Foott 211tr, 435bl, J.&I.Palmer 324c, J.M.Pearson 194bl, Alan Root 189cr, J.Root 211c; **Shell:** 434-5tc; **Harry Smith Collection:** 337cr, 341cl; **Smithsonian Institute, Washington D.C.:** 132br; **Society for Co-operation in Russian & Soviet Studies:** 281bl, br; **Frank Spooner Pictures/Gamma:** Graham 370cla, T.Mackie 314cr, Rotolo 461cl, V.Shone 380clb, 381bc, W.Volz 313br; **Staatlische Museen zu Berlin, PK Antikensammlung/BPK:** 260b; **Standard Fireworks:** 409tl; **Steel Can Recycling Information Bureau:** 419; **Still Pictures:** B.&C. Alexander 367c, N.Dickinson 388c, J.Schytte 438tr; **Tony Stone Images:** 264-5b, 348-9c, 352cr, 355cr, 356cla, cr, 368cl, 377cra, 405bc, 428-9c, 433tc, Glen Allison 5cr, 439tl, Chris Baker 186-7c, P.Berger 389cla, K.Biggs 405tr, Bryn Campbell 349bl, Paul Chesley 348cl, 345cra, 389cl, 462bl, B.Chittock 390tr, P.Correz 468c, R.Frerck 372cla, Roy Giles 465cr, M.Gowan 382c, D.Hanson 354bc, D.Hiser 348bl, 357bc, 391tc, A.Husmo 367b, 404cl, A.Le Garsmeur 382cb, J.Murphy 466bc, S.Proehl 385tl, R.Smyth 463tr, Olaf Soot 357bl, N.Turner 365cr; **Swift Picture Library:** 252-3c; **Sygma:** J.Andanson 371tl, T.H.Barbier-Proken 470br, A.Grace 379cla, A.Gyori 378cl, J.Jones 363cl, A.Nogues 369c.

Telegraph Colour Library: 304cl, 402br, Colorific!/Ken Haas 116tl, /Claus Meyer, Camara Tres 358bl, /Roger Ressmeyer, Wheeler 355tl, Masterfile/H.Blohm 351bl, 353cla, /John Foster 353tl, /J.A.Kraulis 352cla, /V.C.L. 374br; **Topham Picture Source:** 271bc; **Truck Magazine:** 459br, bc.

University of Bristol: Dr.Mervyn Miles 399t; **Jack Vartoogian:** 307tr; **Vauxhall:** 445bl; **Vu Agence:** Christina Garcia Rodero 372cr; **Max Whitaker:** 328cl; **Windsor Castle Royal Library, ©1992 Her Majesty The Queen:** 27tl.

ZEFA: 65bra, c, 182cr, 198bl, 229tr, 289b, 295tc, 312bl, 323cl, 329cra, 335cra, 355bl, 359ca, 364cl, 402tr, 406bl, tr, 438bc, 441c, 459br, 462tl, 465tr, Damm 30, 374cla, Davies 368tr, G.Deichmann/ Transglobe 33b, T.Dimock 399bc, W.Eustep 466bl, J.Feingersch 409br, Freitag 160tr, Goebel 351tl, R.Halin 401tr, Heintgel 168cl, bc, c, Knight & Hunt 33t, Maroon 355crb, W.McIntyre/Allstock 21l, NASA 408tr, R.Nicholas 473tr, Rossenbach 27t, Schlenker 358cla, Schroeter 409tc, B.Simmons 312tl, M.Tortoli 123c, A.Von Humboldt 464cr, T.J.Zhejiang 344bl; **Zinc Galvanising Association:** 443tl.

tl	- top left	clb	- center left below
tr	- top right	crb	- center right below
cla	- center left above	bl	- bottom left
cra	- center right above	br	- bottom right
cr	- center right	tc	- top center
c	- center	cb	- center below

ACKNOWLEDGEMENTS

Photographers
Peter Anderson
Steve Bartholomew
Peter Chadwick
Tina Chambers
Andy Crawford
Colour Company
Geoff Dann
John Downes
Michael Dunning
John Edwards
Lynton Gardiner
Steve Gorton
Colin Keates
Tim Kelly
Gary Kevin
Chris King
Dave King
Cyril Laubscher
Kevin Mallett
Ray Moller
David Murray
Tim Ridley
David Rudkin
Karl Shone
James Stevenson
John Swift
Harry Taylor
Andreas von Einsiedel
Jerry Young

Illustrators
Graham Allen
Norman Barber
David Bergen
Roby Braun
Peter Bull
Joanna Cameron
Jim Channell
Bob Corley
Sandra Doyle
David Fathers
Roy Flooks
Tony Gibbons
Mike Gillah
Tony Graham
Peter Griffiths
Terry Hadler
Edwina Hannah
Charlotte Hard
Kaye Hodges
Keith Hume
Ray Hutchins
Aziz Khan
Pavel Kostal
Norman Lacey
Stuart Lafford
Kenneth Lilly
Linden Artists
Steve Lings
Mike Loates
Chris Lyon
Alan Male
Richard Manning
Janos Marffy
Josephine Martin
Annabel Milne
Sean Milne
Patrick Mulray
Richard Orr
Alex Pang
Darren Pattenden
Liz Pepperell
Jane Pickering
Gill Platt
Maurice Pledger
Sebastian Quigley
Christine Robins
Eric Rome
Michelle Ross
Simon Roulstone
Colin Salmon
John Searl
Pete Serjeant
Rob Shone
Clive Spong
Roger Stewart
John Temperton
Grose Thurston
Graham Turner
Brian Watson
Phil Weare
Sonia Whillock
John Woodcock
Michael Woods

Models
Celia Allen
Roby Braun
Atlas Models
Cheltenham Cutaway Exhibits Ltd.
Crystal Palace Park, London
Arril Johnson
Donks Models
Centaur Studios
Norrie Carr Model Agency
Peter Griffiths
John Holmes
Scallywags Model Agency
Truly Scrumptious Child Model Agency

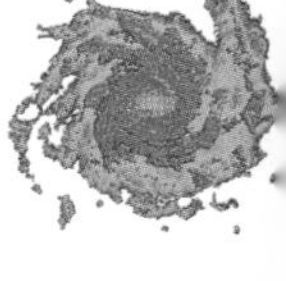